Electrical properties of polymers

Cambridge Solid State Science Series

EDITORS:
Professor R. W. Cahn
Applied Sciences Laboratory, University of Sussex
Professor M. W. Thompson
School of Mathematical and Physical Sciences, University of Sussex
Professor I. M. Ward
Department of Physics, University of Leeds

A. R. Blythe
Senior Research Physicist, ICI Plastics Division

Electrical properties of polymers

CAMBRIDGE UNIVERSITY PRESS
Cambridge
London · New York · Melbourne

Published by the Syndics of the Cambridge University Press
The Pitt Building, Trumpington Street, Cambridge CB2 1RP
Bentley House, 200 Euston Road, London NW1 2DB
32 East 57th Street, New York, NY 10022, USA
296 Beaconsfield Parade, Middle Park, Melbourne 3206, Australia

First published 1979

Printed in Great Britain at
The Alden Press, Oxford

Library of Congress Cataloguing in Publication Data
Blythe, A. R.
Electrical properties of polymers.
(Cambridge solid state science series)
Includes bibliographies and index.
1. Polymers and polymerization –
Electrical properties. I. Title.
QD381.9.E38B56   547′.84   77-85690
ISBN 0 521 21902 7

To Andrew and Michael

Contents

viii   *Contents*

Preface

This book grew out of a series of lectures given to graduate students in the Polymer Physics Group at Leeds University. It is meant as an introductory text on electrical aspects of polymeric materials.

Emphasis is mainly laid on the description and explanation, in molecular and electronic terms, of the observed phenomena, so as to give a basic understanding of the electrical behaviour of polymers. Principles of up-to-date measurement methods are also stressed, since a sound framework of experimental techniques and data is so essential for proper scientific development of such a subject. The choice of subject matter was made with the aim of being educative and stimulating rather than exhaustive and exhausting. It is hoped the book will be of most help to those venturing into research in polymer science or to those joining the plastics industry who want an insight into the somewhat specialised area of electrical properties. Only a general knowledge of physics and chemistry is assumed, and for this reason an introduction to polymer structure is included.

The electrical properties of polymers is a subject which is inherently interdisciplinary in nature, being closely allied with the mechanical properties of polymers on the one hand, and with the semiconductive properties of inorganic substances on the other. A primary objective was to collate the relevant aspects of these contingent subjects to form a more unified treatment than is generally available. Since the early days of plastics technology, when such materials were regarded electrically as simply good insulators, observations of subtleties in electrical response have shed a great deal of light on the underlying microscopic structure and molecular dynamics. This has contributed to polymer science in a general way, and has, at the same time, enabled the development of materials which meet exacting electrical-engineering requirements. The book encompasses both this well-established area of dielectric science as well as the modern frontiers concerned with conductive, even superconductive, plastics. Research along these lines has demonstrated the feasibility of obtaining materials with entirely novel sets of properties.

Electrostatic charging effects are also treated in some detail. This has been made worthwhile by the relatively recent upsurge of activity which has put this old subject on a new scientific footing.

SI units have been used throughout the book.

Special thanks are due to Professor I. M. Ward for encouraging me to write the book in the first place and for reading and commenting on the whole of the draft manuscript. I should also like to thank Dr G. R. Davies for constructive criticism on much of the manuscript and Professor J. S. Dugdale for suggestions on chapter 5. I am very grateful too for the invaluable help and advice from all my colleagues in ICI and for the assistance from the Plastics Division in preparing the typescript and drawings. Mlle H. Bertein, Mr W. Reddish, Dr D. J. Groves and Mr J. Billing kindly supplied photographs.

Finally, I am especially grateful to my wife for her constant moral support during this project.

July 1977   A. R. Blythe

1 Introduction

1.1 Polymers as non-conductive materials

Although unified by direct concern with the effects produced by electric fields, the subject of the electrical properties of polymers covers a diverse range of molecular phenomena. By comparison with metals, where the electrical response is overwhelmingly one of electronic conduction, polymers display a much less striking response. This absence of any overriding conduction does allow, however, a whole set of more subtle electrical effects to be observed more easily. For instance, polarisation resulting from distortion and alignment of molecules under the influence of the applied field becomes apparent. Examination of such polarisation not only gives valuable insight into the nature of the electrical response itself, but it also provides a powerful means of probing molecular dynamics. For this reason electrical studies form a desirable supplement to studies of purely mechanical properties aimed at reaching an understanding of the behaviour of polymers on a molecular basis.

As a consequence of the characteristically insulating nature of solid polymers, any electrostatic charges that they acquire are retained for a long time. Since charges may be deposited by mere contact with a different material, the charged condition is frequently encountered with articles made from polymers. It is worth pointing out here that, although contact charges represent only a slight imbalance of charge compared with the total amounts of positive and negative charge present in matter, they can nevertheless give rise to electric fields which are high enough to cause sparking in air. Thus a surface charge density resulting from just one or two extra electronic charges per million surface atoms is sufficient to generate a field which exceeds the air breakdown value. The occurrence of sparking means that an insulating specimen will often present a complex and confusing distribution of charge on its surface, reflecting a whole history of charging and discharging events. In recent years there has been a resurgence of interest in electrostatics and new experimental techniques have vastly improved the scope for scientific study. The stimulus has come partly from a recognition of the possible benefits to be gained commercially from control and exploitation of static charges on polymeric materials.

As no known insulator is completely free of conduction processes, charges are expected to diffuse away eventually under the influence of their own field, even though this may take many years in extreme cases. Low-level conduction in essentially insulating materials can take a variety of forms. It may very often be attributed to impurities which provide a small concentration of charge carriers in the form of electrons or ions. At high fields the electrodes may also inject new carriers into the polymer, causing the current to increase more rapidly with voltage than in accord with Ohm's law. At very high fields these and other processes, often involving conduction over a surface, inevitably lead to complete failure of the material as an insulator.

1.2   Electrical applications of polymers

The electrical insulating quality inherent in most polymers has long been exploited to constrain and protect currents flowing along chosen paths in conductors and to sustain high electric fields without breaking down. Insulating polymeric materials for early electrical equipment were made from naturally occurring products. For example, the first trans-Atlantic telephone cables laid in the 1860s were insulated with gutta-percha, which is one of the polymers extracted from rubber trees. As synthetic high polymers became available in the twentieth century, the range of insulators was continually improved. The great virtue of these new materials, such as polystyrene, was their combination of high quality of insulation with ease of fabrication by moulding. Polyethylene, which combines superb insulating properties with mouldability and a high degree of toughness and flexibility, arrived on the scene just in time for the more demanding applications of insulation in coaxial cables for radar apparatus and television. More recently extreme requirements for very low-conductivity materials, used in electret microphones, for example, have been met by fluorinated polymers. High-performance thin films have also been developed for various types of capacitors.

The choice of material for a particular application naturally depends on being able to reach a good compromise amongst a whole range of considerations, including mechanical properties, ease of fabrication into a final product, and cost. The basic insulating properties of polymers are more than adequate for many purposes, and any development effort may then be primarily concentrated on improving other aspects of the material's performance. High on the list will be a need for good chemical and physical stability in the working environment, which might involve exposure to strong sunlight, organic solvents and high temperatures. Commercial success for a polymeric material for certain highly specialised electrical applications will, on the other hand, depend on very careful *grooming* with respect to the most demanding electrical quality expected of it. Thus some applications will require that every precaution be taken to eliminate vestigial DC conduction effects or to reduce variation of polarisation effects with AC frequency to an absolute minimum. In order to meet the specifications for high-voltage use, priority must be given to the need for a high breakdown strength which can be maintained in the face of practical environmental conditions. Only on the basis of detailed knowledge and understanding of the molecular structure and behaviour of the basic polymers can one hope to approach the optimum in performance. This has motivated the investigation in depth of the electrical behaviour of many polymeric systems.

Polymers are almost always good insulators, but that is not to say that a conducting plastic is not desirable! A lightweight, readily mouldable, highly conductive material has long been recognised as a worthwhile goal to strive for, and considerable scientific research has been devoted to this. Although encouraging results have been obtained, there is still a long way to go in developing useful products. Apart from the obvious market for a highly conductive material which would be suitable for power and signal transmission, there is also one for materials having intermediate conductive properties, e.g. for flexible heating elements and graded cable insulations. Certain of these can be met to some extent by modifications to existing polymers. Thus conductive composites made by mixing graphitic powder into the bulk of a polymer are commonly used. Static charges are generally an embarrassment on manufactured articles and materials – they attract dust, cause film to cling to metal surfaces and can give electric shocks to people. There is real benefit to be gained from a slight degree of conduction which allows charges to leak away to earth. For this purpose surface treatments with so-called antistatic agents are widely used on ordinary plastics to give sufficient conduction over the surface without destroying other valuable properties of the material. Unfortunately, this kind of solution is not entirely satisfactory, because the treatment tends to wear off; an inherent, bulk conduction would be much superior.

1.3   Structure of polymers

This section outlines the main features of polymer structure, as a basis for subsequent explanation and discussion of electrical behaviour. Most of the synthetic high polymers with which this book is concerned are organic compounds consisting of long, chain-like molecules where repeated molecular units are linked by covalent bonds. A single molecular chain commonly contains a thousand or more repeat units and reaches a total length in excess of 1 μm. The atoms which form the *backbone* of organic polymers are predominantly carbon, sometimes in combination with oxygen and/or nitrogen.

1.3.1   Chemical structure

The process of polymerisation in which small molecules of the starting material, the monomer, undergo chemical reaction together to form long chains may proceed in a variety of ways. These may be divided into two principal categories, *addition* and *condensation* polymerisation, which have major structural implications for the final product.

Addition polymerisation is typified by the conversion of ethylene to polyethylene according to the following equation:

$$n\text{CH}_2{=}\text{CH}_2 \longrightarrow (-\text{CH}_2-\text{CH}_2-)_n$$

In this joining together of molecules of ethylene the necessary linking bonds are obtained by breaking the original double bonds, which are relatively unstable. The chain reaction may be initiated with the aid of a catalyst, often a peroxide, which provides a chemically reactive species to attack a limited number of double bonds. Sequential addition of ethylene molecules then proceeds until all the ethylene is used up. None of the atoms of the original monomer, ethylene in this case, is excluded from the final polymer, the addition process only involving rearrangement of bonds.

Addition polymers include the important class derived from vinyl monomers, general formula $\text{CH}_2{=}\text{CHX}$, having a backbone consisting entirely of carbon atoms. Some examples of vinyl polymers and closely related ones are given in table 1.1.

Polyethers, which contain oxygen as well as carbon atoms in their backbones, are examples of addition polymers that are generally made by a ring-opening reaction. Thus the cyclic compound ethylene oxide polymerises to a polyether as follows:

$$n\overset{\displaystyle\text{O}}{\overset{\displaystyle\diagup\diagdown}{\text{CH}_2-\text{CH}_2}} \longrightarrow (-\text{CH}_2-\text{CH}_2-\text{O}-)_n$$

Condensation polymerisation occurs when multifunctional monomers, i.e. ones possessing more than one chemically reactive group per molecule, react together with the elimination of small molecules, usually water. With a bifunctional monomer the product is a linear

Table 1.1. *List of vinyl and related polymers*

Monomer formula	Polymer name
$CH_2{=}CH_2$	Polyethylene
$CH_2{=}\underset{\underset{CH_3}{\mid}}{CH}$	Polypropylene
$CH_2{=}\underset{\underset{Cl}{\mid}}{CH}$	Poly(vinyl chloride)
$CH_2{=}\underset{\underset{C_6H_5}{\mid}}{CH}$	Polystyrene
$CH_2{=}\underset{\underset{COOCH_3}{\mid}}{C}{-}CH_3$	Poly(methyl methacrylate)
$CF_2{=}CF_2$	Polytetrafluoroethylene

polymer, e.g. the polyamide Nylon-6 is derived from ε-amino caproic acid as follows:

$$NH_2(CH_2)_5CO\overline{OH + H}NH(CH_2)_5COOH \longrightarrow$$
$$NH_2(CH_2)_5CO \cdot NH(CH_2)_5COOH + H_2O \text{ etc.}$$

Quite often polymerisation proceeds by interaction of pairs of complementary monomers. Thus Nylon-6,6 is formed by reaction of adipic acid with hexamethylene diamine:

$$HOOC(CH_2)_4COOH + NH_2(CH_2)_6 \cdots \longrightarrow$$
$$\longrightarrow HOOC(CH_2)CO \cdot \cdots \qquad H_2O \text{ etc.}$$

The formation of poly(ethylene terephthala terephthalic acid is another very well-knov

$$nHO(CH_2)_2OH + nHOOC\langle O \rangle C$$
$$HO[(CH_2)_2$$

Some common condensation polyme

The final product of polymerisation tion type, will contain chains coverin

Table 1.2. *List of condensation polymers*

Polymer name	Chemical formula
Polyester (Terylene, Dacron)	$[-(CH_2)_2O \cdot CO \langle \bigcirc \rangle CO \cdot O-]_n$
Polyamide (Nylon-6,6)	$[-(NH(CH_2)_6NH \cdot CO(CH_2)_4 \cdot CO-]_n$
Polycarbonate	$[-\bigcirc-\overset{\overset{CH_3}{\vert}}{\underset{\underset{CH_3}{\vert}}{C}}-\bigcirc-O-\overset{\overset{O}{\parallel}}{C}-O-]_n$
Polyethersulphone	$[-\bigcirc-O-\bigcirc-\overset{\overset{O}{\parallel}}{\underset{\underset{O}{\parallel}}{S}}-]_n$

possible to specify an average molecular weight. The number-average molecular weight $\bar{M}_n$ is defined by

$$\bar{M}_n = \frac{\Sigma M_x n_x}{\Sigma n_x}$$

where M_x and n_x are the molecular weight and relative number, respectively, of the species containing x repeat units.

1.3.2 Chemical variants

Many variations from linear, regular chain formation may occur during polymerisation. The first set of variations arises from the presence of alternative reactive centres either on the monomer or on the already formed polymer chains. Thus the uniformity of *head-to-tail* monomer reaction

$$-CH_2-\underset{\underset{X}{\vert}}{CH}-CH_2-\underset{\underset{X}{\vert}}{CH}-CH_2-\underset{\underset{X}{\vert}}{CH}-$$

may be broken, and some *head-to-head* and *tail-to-tail* structures incorporated in the chain:

$$-CH_2-\underset{\underset{X}{\vert}}{CH}-\underset{\underset{X}{\vert}}{CH}-CH_2-CH_2-\underset{\underset{X}{\vert}}{CH}-\underset{\underset{X}{\vert}}{CH}-$$

f one particular point in the monomer usually predominates, so that the effect is relatively small. A 1% head-to-head content may, however, seriously upset crystallisation.

Some polymerisations do not continue in a linear fashion, and *branching* occurs, as shown in fig. 1.1(*a*). When branching is prevalent, it can have a serious effect on properties. Thus the polymerisation of ethylene under high-pressure conditions gives a product which has so many side chains on each main chain that crystallisation is appreciably suppressed. This material is softer than the highly crystalline linear form from a

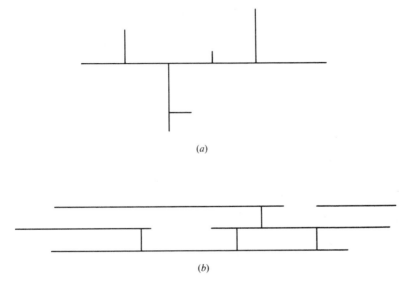

(*a*)

(*b*)

Fig. 1.1. Diagrams of variations from linear polymerisation: (*a*) branching, (*b*) crosslinking.

catalytic low-pressure process. The two forms of polyethylene may be distinguished by a difference in density between them, the more crystalline material being the more dense.

Polymer chains may also suffer a degree of *crosslinking*, as shown in fig. 1.1(*b*), either during the initial polymerisation, or at some subsequent stage. In a condensation polymerisation where the monomer molecules each possess three reactive groups, i.e. are trifunctional, crosslinking is inevitably a major feature throughout the polymerisation. Furthermore, it may be formally shown that once two-thirds of the total number of reactive groups in such a system have reacted, the polymer will be rapidly combined into one giant, crosslinked molecule or *gel*. The onset of this 3-dimensional network formation will be reflected in a sudden increase in molecular weight to a very large value over a small extent of reaction. When extensive crosslinking is promoted, as in *thermosetting* materials, the product becomes typically a hard, brittle resin. For some purposes crosslinks are often deliberately introduced into a linear polymer by a post-curing process, e.g. by heating with peroxides. In the rubber industry light crosslinking with sulphur (vulcanisation) is used to convert the starting polymers (natural or synthetic latex) into a tough product with a high extensive elasticity.

As in any other chemical compound, different geometrical arrangements of substituent groups are possible in a polymer where rigid molecular units are involved. This gives rise to *cis*- and *trans-configurational isomerism* in polymers containing double bonds in their repeat units, as in natural or synthetic rubbers:

cis-isomer   trans-isomer

Such differences in structure can have a profound effect on the physical properties of a polymer. Thus natural rubber, which comprises *cis*-1,4-polyisoprene, is a soft rubbery material at room temperature, whereas gutta-percha, which comprises the corresponding *trans*-isomer, is semicrystalline and hard. The method of polymerisation determines the isomeric form of the polymer.

When the repeat units of a polymer chain are themselves asymmetric, on account of their containing a carbon atom with four different substituents, the large number of possible permutations of right-handed (d) and left-handed (l) units represents a large number of *steric*-isomers. This kind of isomerism in polymers is called *tacticity*. When the arrangement along the chain is completely ordered, the polymer is said to be *stereoregular*. In a substituted vinyl polymer like polypropylene, $-[CH_2-CH(CH_3)-]_n$, there are three main types of steric isomer:

(*a*) *Isotactic*. In this isomer all repeat units are identical, either right-handed (d) or left-handed (l) forms, and arranged in a linear, head-to-tail fashion. In the fully extended (planar zigzag) chain all the methyl substituents lie on the same side of the chain (fig. 1.2(*a*)).

(*b*) *Syndiotactic*. Here successive repeat units alternate in configuration, *dldldl . . .*, and successive methyl groups along the fully extended chain lie on alternate sides (fig. 1.2(*b*)). More elaborate ordering does not often occur in practice.

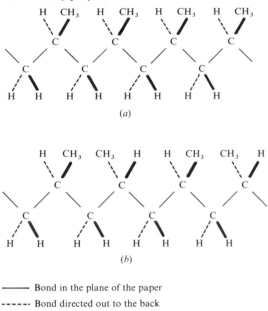

(a)

(b)

——— Bond in the plane of the paper
----- Bond directed out to the back
▬▬▬ Bond directed out to the front

Fig. 1.2. Stereo-isomers of polypropylene: (a) isotactic, (b) syndiotactic.

(c) *Atactic*. This label is applied to completely random arrangements of *d* and *l* units along the chain.

A high degree of stereoregularity usually increases crystallinity, leading to enhanced rigidity. For this reason certain catalysts which favour stereoregularity, e.g. Natta–Ziegler catalysts for propylene polymerisation, are of great commercial importance.

1.3.3   Conformation and hindered rotation

Rotation about a single, covalent bond between two carbon atoms is only restricted by the spatial interaction of the other atoms or groups attached to the two carbons. The molecular arrangements that are thereby allowed are called *conformational* isomers. The principles and nomenclature of conformation can be most easily described in the case of a simple organic molecule like n-butane, and the situation is illustrated in fig. 1.3. From the potential energy curve with respect to the angle ϕ of rotation around the central C–C bond we can readily appreciate that the highest energy occurs in the fully *eclipsed* state ($\phi = 180°$), where the terminal methyl groups most strongly interact with each

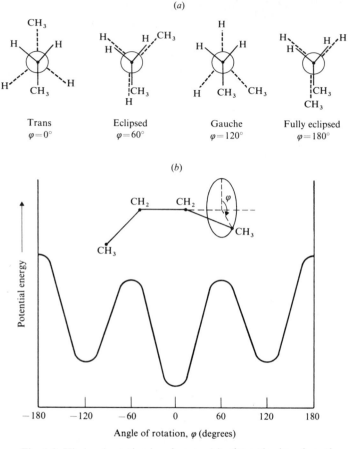

Fig. 1.3. Hindered rotation in n-butane: (*a*) schematic view along the central C–C bond, showing various conformations, (*b*) potential-energy curve for rotation about the central C–C bond.

other. Conversely, the most stable state corresponds to the *trans*-conformation ($\phi = 0°$) whose energy is about 15 kJ mol^{-1} lower than that of the eclipsed form. Two other staggered states, called *gauche*-conformations ($\phi = \pm 120°$), are also relatively stable. In a carbon-chain polymer the rotational arrangement about the whole sequence of bonds along the backbone determines the shape or conformation of the molecule.

Since the energy barriers to internal rotation are rather low, typically not much larger than thermal energies at room temperature, polymer chains may usually reorganise themselves rapidly under an applied stress at modest temperatures. This, in turn, endows an uncrosslinked or *thermoplastic* polymer with pronounced flexibility and toughness.

Natural or synthetic rubbers owe their extreme pliability partly to extremely low rotational barriers about single bonds; although the double bonds that are present along the chain are themselves rigid, they impose reduced steric hindrance upon rotation about adjacent single bonds. At elevated temperatures thermal energies are quite sufficient for most rotational barriers to be rapidly overcome and this is the basis for one of the most useful features of thermoplastics – they may be softened and re-formed by the application of heat.

1.3.4   Copolymers

Copolymers, in which individual molecular chains contain a mixture of two or more different repeat units, are often encountered in practice. The compositional pattern of copolymers depends on the conditions of polymerisation and may be either *random,*

–A—B—B—A—B—A—A–

or divided into *blocks* containing long sequences of like units:

–[A]—[B]$_m$—[A]$_n$–

By special methods it is also possible to produce *graft* copolymers where branches of a different polymer are tagged on to a polymer backbone:

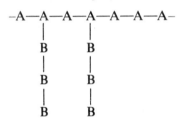

Also chains may be subsequently *extended* with a different polymer:

B—B—A—A—A—A—A—A—A—B—B

In random copolymers chain flexibility and crystallinity can be quite different from those in either of the component *homo*polymers. In consequence the copolymer may well present an entirely new set of physical properties. On the other hand the different chain segments in block or graft copolymers often segregate into effectively separate phases, when the properties of the composite resemble those of a mixture of the individual homopolymers.

1.3.5   Crystallisation and orientation

Long, flexible molecules naturally become highly contorted and entangled with each other in the molten or dissolved state, and reorganisation into a regular, crystalline array is then an awkward process. If a molten polymer is rapidly cooled or quenched there is insufficient time for any crystallisation to occur, and a transparent, amorphous solid is obtained. This apart, at least some crystallisation will usually take place in most polymers, and the typical milky appearance of a polymer is indicative of a semicrystalline state consisting of a mixture of small crystalline regions (crystallites) and amorphous matter. The exact degree of crystallinity, most often in the range 5 to 50%, depends on the inherent ease of crystallisation of the polymer molecules (as mentioned previously, factors like branching, crosslinking, stereo-irregularity and copolymerisation can suppress crystallisation) and on the thermomechanical history of the particular sample under observation.

The amorphous or disordered phase of a polymer normally shows a transition from a rubbery to a glassy condition at some characteristic transition temperature T_g. Crystallisation can only proceed above the glass transition temperature where sufficient molecular mobility is present for reorganisation to occur. The glass transition is a difficult phenomenon to explain in detail and the subject is developed further in §3.3.

When a polymer is stretched or drawn, the internal shearing action tends to align the long molecules preferentially in the stretch direction, and this orientation can be detected by optical methods. Not surprisingly such 1-dimensional ordering tends to induce crystallisation. In distinctly non-crystallisable polymers like commercial polystyrene and poly(methyl methacrylate), made by free-radical initiation at high temperature, 3-dimensional order is still absent after stretching, however, and the resulting structure must be regarded as a somewhat elongated tangled skein in these cases.

The extra strength that orientation of chain molecules imparts to a material in the direction of orientation (at the expense of strength in the other directions) is widely exploited in the commercial production of films and fibres. Typically these materials are made by subjecting the polymer to a high degree of drawing. The resulting orientation is *set* by crystallising whilst the polymer is still under tension above T_g.

The structure of truly crystalline regions of a polymer may be determined from wide-angle X-ray diffraction patterns. In this way the unit cell dimensions and molecular configurations have been obtained for all the common polymers. In general the molecules of the crystalline phase adopt an extended zigzag or helical conformation and lie parallel to each other.

By contrast, total molecular architecture, involving the conjunction

of crystalline and amorphous parts, has proved to be much less amen-
able to investigation. For a long time structural interpretations were
based on the *fringed-micelle* model in which molecules are supposed to
wander through ordered and disordered regions, as shown in fig. 1.4(*a*).
More recently views have been revised in the light of the discovery of a
specific mode of crystallisation of single crystals from dilute solution.
There, the principal crystal habit, well exemplified by linear polyethy-

(*a*)

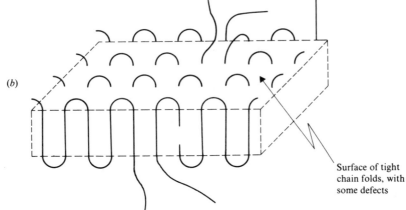

(*b*)

Surface of tight
chain folds, with
some defects

Fig. 1.4. Diagrams of polymer crystallinity models: (*a*) fringed micelles, (*b*) a
chain-folded lamella.

lene, is a *lamella*, with lateral dimensions of the order of 10–20 μm and with a thickness of only about 10 nm. Electron-diffraction studies show, surprisingly, that the chain axis (usually denoted the *c*-axis) of the polymer molecules lies approximately perpendicular to the plane of the lamella, even though the lamella thickness is much shorter than the chain. This may be explained by the chains folding back on themselves in order to fit as shown in fig. 1.4(*b*). The exact arrangement of the folds is difficult to characterise unambiguously, but in some cases it is believed to be tight, with chains re-entering the crystal at adjacent sites. It is now generally held that chain-folding is also a common feature in crystallisation from the melt.

When a quiescent melt is cooled the gross features of crystallisation are governed by the nucleation process. Microcrystalline structures, called spherulites, grow radially from nucleation centres, and the ultimate size of the spherulite will be determined by the original concentration of nuclei. In a very clean and pure polymer where suitable nuclei are scarce, the spherulites may grow large enough to be seen with the naked eye. The main structure of a spherulite is a radial array of twisted lamella-like blades called fibrils. Although some secondary crystallisation may take place later between the fibrils, considerable amounts of non-crystalline material are left there, and between adjacent spherulites. The orientation produced by any mechanical shearing in a melt will modify the course of a crystallisation process, and the final structure will possess a degree of anisotropy.

The interplay of orientation and crystallisation leads to a wide range of supermolecular structures or *morphologies*. Each different morphology represents to the user a different compromise in physical properties, so that characterisation and control of morphology becomes very important for the efficient application of polymeric materials.

1.4   Further reading

Many basic texts on polymer science are available. Wide coverage is given in the book by Billmeyer (1971). Stille (1962) provides a useful introduction to polymer chemistry.

2   Dielectrics in static fields

2.1   Electrostatic relations

The degree to which a material responds to an applied electric field can be most easily appreciated in the case of a parallel-plate capacitor. Suppose first that a fixed voltage V is connected across such a capacitor where the plates are separated by a distance d in a vacuum, as shown in fig. 2.1(a). Neglecting edge effects, the electric field E thereby produced in the region between the plates will be uniform, having a magnitude

$$E = V/d. \tag{2.1}$$

We should note that since an electric field has direction as well as magnitude it is more fully represented as a vector quantity denoted by E, with components E_x, E_y, E_z. In the present case, where the field is necessarily perpendicular to the plates, it follows from Coulomb's law* that the charges $+Q$ and $-Q$ per unit area stored on the plates are directly proportional to the magnitude of the field, i.e.

$$Q = \varepsilon_0 E. \tag{2.2}$$

The constant of proportionality ε_0 is called the permittivity of free space and has the value 8.85×10^{-12} F m^{-1}. The vacuum capacitance per unit area of electrode C_0 is defined as the ratio of the stored charge per unit electrode area to the applied voltage, i.e.

$$C_0 = Q/V. \tag{2.3}$$

Now consider the capacitor with the material of interest between its plates, as shown in fig. 2.1(b). The material will respond to the applied electric field by redistributing its component charges (electrons and protons) to some extent, positive charges being attracted towards the

---

* In SI units Coulomb's fundamental inverse-square law governing the force F between two point charges q_1 and q_2 separated by the vector distance r in a vacuum takes the form:

$$F = \frac{q_1 q_2 r}{4\pi \, \varepsilon_0 r^3},$$

where ε_0 is the permittivity of free space. The factor 4π is present in the denominator of the above formula because permittivity (along with permeability) is rationalised, in order to avoid the continual appearance of the factor 4π in Maxwell's electromagnetic field equations.

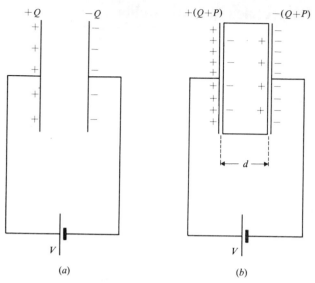

Fig. 2.1. Charges on a parallel-plate capacitor with (*a*) a vacuum between the plates, and (*b*) a dielectric between the plates.

negative electrode and vice versa. This effect is called polarisation of the material. If the material is isotropic, the effect in the uniform field of the parallel plate capacitor will amount to the production of a small dipole moment $P\mathrm{d}v$ aligned in the field direction in each elemental volume dv. The polarisation P is thus defined as a vector quantity expressing the magnitude and direction of the electric moment per unit volume induced in the material by the applied field. Unless the field is very high, the magnitude of the polarisation is directly proportional to the field. (In the case of an anisotropic material its direction is not necessarily parallel to the field.) Each volume element of a polarised material will behave electrically as though it consisted of charges $+q$ and $-q$ separated by a distance l in the field direction, giving a dipole moment $ql = P\mathrm{d}v$. These dipoles will combine in an analogous manner to the head-to-tail addi-tion of magnets, to produce charges $+P$ and $-P$ per unit area on the surfaces adjacent to the electrodes. Green's theorem may be used to show rigorously that the field due to a uniform polarisation P in a sample is equivalent to that from a distribution of charge P_n on the bounding surface where P_n is the normal component of the polarisation at the surface. The presence of these polarisation or *bound* charges means that more charge can be stored on the capacitor electrodes for the same applied voltage and the capacitance of the system is thereby increased. The ratio ε of the increased capacitance to the vacuum capacitance,

$$\varepsilon = \frac{C}{C_0} = \frac{Q+P}{Q}, \tag{2.4}$$

varies from material to material depending on the amount of polarisation which occurs in the material. This characteristic ratio is essentially independent of the applied voltage and is therefore also independent of the electric field; it is commonly called the dielectric constant of the material. Substituting for Q from equation (2.2) in equation (2.4), and adopting the more general vector notation,

$$\varepsilon = \frac{\varepsilon_0 E + P}{\varepsilon_0 E}, \tag{2.5}$$

or

$$P = (\varepsilon - 1)\varepsilon_0 E.$$

The quantity $\varepsilon_0 \varepsilon E$, called the electric displacement D in the material, may be obtained by rearranging equation (2.5):

$$D = \varepsilon_0 \varepsilon E = \varepsilon_0 E + P. \tag{2.6}$$

This is the fundamental electric field equation which applies at any point in an isotropic medium. In this context the quantity $\varepsilon_0 \varepsilon$ is the absolute permittivity of the material, and the ratio ε which we have called the dielectric constant of the material is more properly termed the relative permittivity (with respect to the absolute permittivity of free space ε_0). The flux of dielectric displacement begins and ends on free charge and otherwise is continuous, even at an interface between two media. Electric field on the other hand is discontinuous at an interface between two different materials as a result of the different degrees of polarisation.

If we think of polarisation on the molecular level, we can say that the effect of the applied electric field is to induce an electric dipole m on each individual molecule, the magnitude depending on the local electric field strength E_L at the molecule:

$$m = \alpha E_L. \tag{2.7}$$

The constant of proportionality α is called the polarisability of the molecule. Except in certain anisotropic cases, the average direction of the induced molecular dipoles is in the direction of the applied field, and since the local field is, as we shall see later, proportional to the overall applied field, we have $m \propto E$. The total dipole moment per unit volume developed in this way, the polarisation P, is then related to the number of molecules per unit volume N_0:

$$P = N_0 \alpha E_L. \tag{2.8}$$

2.2   Molecular polarisability

We must now look in more detail at polarisation at the molecular or microscopic level. There are three components of molecular polarisation:

(a) *Electronic polarisation.* An electric field will cause a slight displacement of the electrons of any atom with respect to the positive nucleus. The shift is quite small because the applied electric field is usually quite weak relative to the intra-atomic field at an electron due to the nucleus. Thus taking the charge on a proton to be 1.6×10^{-19} C and a typical atomic radius to be 10^{-10} m, the electric field on an electron will be of the order of 10^{11} V m^{-1}, whereas externally applied fields seldom exceed 10^8 V m^{-1}. Electronic polarisation can react, however, to very high frequencies and is responsible for the refraction of light.

(b) *Atomic polarisation.* An electric field can also distort the arrangement of atomic nuclei in a molecule or lattice. The movement of heavy nuclei is more sluggish than electrons so that atomic polarisation cannot occur at such high frequencies as electronic polarisation, and it is not observed above infra-red frequencies. Now in the case of molecular solids we know from vibrational spectroscopy that the force constants for bending or twisting of molecules, involving changes in angles between bonds, are generally much lower than those for bond stretching, so we may expect bending modes to make the major contribution to atomic polarisation. The magnitude of atomic polarisation is usually quite small, often only one-tenth of that of electronic polarisation, although there are exceptions where a particular mode of bending produces relatively large departures from the normally symmetric arrangements of positive and negative centres within the molecule. In ionic compounds the effect can sometimes be very large. For instance, there is a special contribution in the case of crystalline sodium chloride from a relative shift of all the positive ions with respect to the negative ones.

(c) *Orientational polarisation.* If the molecules already possess permanent dipole moments, there is a tendency for these to be aligned by the applied field to give a net polarisation in that direction. This effect is discussed in the next section, and the rate of dipolar orientation, which is highly dependent on molecule–molecule interaction, is the subject of the next chapter. Suffice it to say at this stage that orientation of molecular dipoles can make a contribution which is large, but which may be slow to develop, to the total polarisation of a material in an applied field.

Fig. 2.2 depicts the characteristic stepwise fall in polarisation of a material as the measurement frequency is raised, rendering it impossible for successive components of molecular polarisation to make their contribution. The dielectric constant follows a similar pattern.

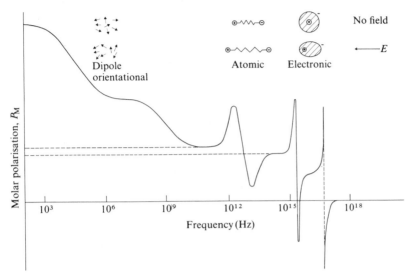

Fig. 2.2. Dispersion of molar polarisation in a dielectric (schematic).

2.3   The local field

As soon as we consider the molecular nature of a material, we realise that the internal electric field will vary from point to point as a consequence of the interaction of fields from the dipoles which are induced on each molecule by the applied field, although the space-average electric field over a volume large in comparison with molecular size (this is equivalent to the classical electric field based on a continuum model) may still be uniform. The field acting on an individual polarisable entity like an atom or molecule is called the local field E_L, and it is an important concept in linking observable bulk behaviour of a material with the properties of its constituent atoms or molecules.

In calculating a local field it is convenient to divide it into two principal components:

$$E_L = E_C + \Sigma E_D, \tag{2.9}$$

where E_C is the field due to the real charges on the electrodes which sustain the externally applied field, and ΣE_D is the sum of fields from all the molecular dipoles apart from that on the molecule at the reference point within the sample. A general evaluation is difficult because the sum ΣE_D does not converge for an infinitely large sample and it depends on the shape of the sample. An approach which was first adopted by Lorentz and which has been of great value in the development of dielectric theory is based on the following simple model. Consider a sample of material contained between the electrodes of a parallel-plate capacitor, as shown in fig. 2.3, and suppose that an imaginary sphere is

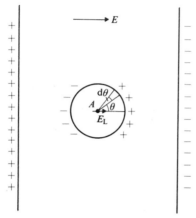

Fig. 2.3. Model for the Lorentz local field.

drawn round a particular molecule at point A. The local field at A may then be written as

$$E_L = E_C + E_P + E_M,\qquad(2.10)$$

where ΣE_D has been subdivided into a component E_P from all the sample which lies outside the sphere and a component E_M from that inside the sphere. If the radius r of the sphere is chosen to be small in comparison with the size of capacitor, but large in comparison with molecular dimensions, it is permissible to calculate E_P on the basis of classical electrostatic theory assuming that the material outside the sphere is a continuum with a dielectric constant ε. Discrete molecular structure need only be taken into account in calculating E_M. Let us now consider the separate components of the local field.

Since the total charge per unit area on the electrodes as shown previously is $(Q+P)$, we have directly for the magnitude of the field from this source

$$E_C = \frac{Q+P}{\varepsilon_0}. \tag{2.11}$$

Substituting for Q from equation (2.2) we obtain

$$E_C = E + \frac{P}{\varepsilon_0}, \tag{2.12}$$

where E is the applied field $(E = V/d)$.

The field at A from the polarisation of the material outside the sphere may be further resolved into two parts in terms of the apparent surface charges which it produces at its boundaries. The first part is due to the apparent charges $-P$ and $+P$ per unit area at the surface adjacent to the electrodes and has the value $-P/\varepsilon_0$. The second part is due to the apparent charges $P\cos\theta$ per unit area on the surface of the sphere, where θ is the angle shown in fig. 2.3, and it may be evaluated by integrating over surface elements confined between angles θ and $(\theta+d\theta)$. Such an element has an area $2\pi r^2 \sin\theta\, d\theta$. By symmetry the fields perpendicular to P at the centre of the sphere cancel out and the net field in the direction parallel to P takes the value

$$\int_0^\pi \frac{2\pi r^2 \sin\theta\, d\theta\, P\cos\theta\cos\theta}{4\pi\, \varepsilon_0 r^2} = \frac{P}{3\varepsilon_0}. \tag{2.13}$$

Hence we have

$$E_P = -\frac{P}{\varepsilon_0} + \frac{P}{3\varepsilon_0}. \tag{2.14}$$

E_M, the field at A due to the molecules inside the sphere, will depend on the detailed arrangement of the molecules. For certain special cases, including simple cubic lattices and completely random arrays,

$$E_M = 0. \tag{2.15}$$

In these cases we finally obtain for the local field at A

$$E_L = E + \frac{P}{\varepsilon_0} - \frac{P}{\varepsilon_0} + \frac{P}{3\varepsilon_0} + 0, \tag{2.16}$$

i.e.

$$E_L = E + \frac{P}{3\varepsilon_0}.$$

Substituting for P from equation (2.5),

$$E_L = \frac{\varepsilon + 2}{3} E. \tag{2.17}$$

2.4   The Clausius–Mosotti relation

Knowing the local field operating at each molecule from equation (2.17) above, we may now calculate the individual contributions to the polarisation from equation (2.7):

$$m = \alpha \frac{\varepsilon + 2}{3} E. \tag{2.18}$$

The total polarisation is given by substitution in equation (2.8):

$$P = N_0 \alpha \frac{\varepsilon + 2}{3} E. \tag{2.19}$$

Again substituting for P from equation (2.5), we obtain the Clausius–Mosotti relation:

$$\frac{\varepsilon - 1}{\varepsilon + 2} = \frac{N_0 \alpha}{3\varepsilon_0}. \tag{2.20}$$

If M_w is the molecular weight of the material and ρ is its density this equation may be written

$$\frac{\varepsilon - 1}{\varepsilon + 2} \frac{M_w}{\rho} = \frac{N_A \alpha}{3\varepsilon_0}, \tag{2.21}$$

where N_A is Avogadro's number. The quantity $N_A \alpha / 3\varepsilon_0$ is called the molar polarisation P_M, and it has the dimensions of a volume. So long as the foregoing theory remains valid, the molar polarisation will be constant for a particular material irrespective of the temperature or pressure. We might expect this to be true for a gas or vapour where the density is low, so that intermolecular interaction plays only a minor role; the approximation $E_M = 0$ is then valid. We must also remember that we have not yet taken into account the effect of any orientation of molecular dipoles in the applied field. Fortunately, this latter complication may be circumvented altogether by using the high-frequency value of the dielectric constant which is obtainable from Maxwell's identity between dielectric constant and the square of the refractive index of light n:

$$\varepsilon = n^2. \tag{2.22}$$

Orientation of molecules is much too slow to contribute to polarisation at these high frequencies. Then substituting (2.22) in (2.21) gives a quantity that is usually called the molar refraction of the material:

$$R_M = \frac{n^2 - 1}{n^2 + 2} \frac{M_w}{\rho} = \frac{N_A \alpha}{3\varepsilon_0}. \tag{2.23}$$

Equation (2.23), known as the Lorentz–Lorenz relation, provides a method of calculating the molecular polarisability from a macroscopic, observable quantity, the refractive index. We must make the proviso that we stay away from any resonant absorption frequency, where the refractive index is anomalously high. If the refractive index refers to optical frequencies, the polarisability α will be purely electronic in origin. In practice, electronic polarisabilities derived in this way are remarkably insensitive to temperature and pressure, even for highly condensed phases in which intermolecular forces must be large. This is illustrated for the particular case of xenon in table 2.1.

Table 2.1. *Molecular polarisability of xenon*

State	Polarisability ($\times 10^{-40}$ F m^2)
Gas at room temperature and pressure	4.45
Liquid at the b.p.	4.27
Liquid at the m.p.	4.26
Solid at the m.p.	4.32

A useful property of molar refraction is its additivity. Thus to a first approximation the molar refraction of a given molecule may be predicted by taking the sum of contributions from its various parts. By applying this principle to a large number of compounds, a self-consistent set of polarisabilities may be derived for a wide range of atoms, ions and molecules. For organic compounds a particularly successful approach has been to regard the molar refraction of a symmetrical hydrocarbon like methane as the sum of four equal C–H bond refractions. On this kind of basis a self-consistent set of bond refractions has been established, and a short list (Vogel *et al.*, 1952) is given in table 2.2. It is then a simple matter to calculate the molar refraction of any given compound by adding up the values for all its component bonds. Thus the molar refraction for butan-2-one, $CH_3 \cdot CO \cdot CH_2 \cdot CH_3$, would be calculated as 20.66 cm^3. The observed value is 20.79 cm^3 which is in line with the 1% agreement that is usually obtained. The method works well

Table 2.2. *Bond refractions for the sodium D-line*

Bond	Refraction ($\times 10^{-6}$ m^3)
C—H	1.676
C—C	1.296
C=C	4.17
C≡C (terminal)	5.87
C≡C (non-terminal)	6.24
C—C (aromatic)	2.688
C—F	1.44
C—Cl	6.51
C—Br	9.39
C—I	14.61
C—O (ether)	1.54
C=O	3.32
C—S	4.61
C—N	1.54
C=N	3.76
C≡N	4.82
O—H (alcohol)	1.66
O—H (acid)	1.80
N—H	1.76

because it inherently takes into account the different polarisabilities of bonding electrons. In the case of a high polymer, e.g. polyethylene, $CH_3-(CH_2)_n-CH_3$, one can ignore the chain ends and simply calculate a molar refraction of the repeat unit ($-CH_2-$). The system fails for molecules which contain conjugated double bonds where electrons are delocalised over many atoms. Exceptionally high molar refractions are observed for these molecules; the effect is called optical exaltation.

2.5   Polar molecules

In contrast to molar polarisation calculated from optical refractivities, that calculated from dielectric constants observed at lower frequencies is by no means always independent of temperature. Actually, materials tend to fall into one of two classes. Those in one class show a relatively constant molar polarisation in accord with the simple Clausius–Mosotti relation, whilst the members of the other class, which contains materials with relatively high dielectric constants, show a molar polarisation that decreases with increase in temperature. Debye recognised that permanent molecular dipole moments were responsible for the anomalous behaviour. From theories of chemical bonding we know that certain molecules which combine atoms of different electronegativity are partially ionic and consequently have a permanent dipole moment. Thus chlorine is highly electronegative and the carbon–chlorine bond consequently entails an excess negative charge near the chlorine atom, leaving the carbon with a net positive charge:

$$
\overset{+}{C}\,\text{—}\,\overset{-}{Cl}
$$

The order of magnitude of such a dipole will be equal to the product of the charge on an electron (1.6×10^{-19} C) and a typical bond length (10^{-10} m) giving a value of $\sim 10^{-29}$ C m. The old electrostatic Debye unit of molecular dipoles was in fact equivalent to 3.335×10^{-30} C m. Orientation of such molecular dipoles will clearly produce an extra contribution to the molar polarisation in addition to dipoles *induced* by the applied field, and it is not unreasonable to expect that the equilibrium degree of orientation in a given field will depend on temperature.

First consider the application of an external electric field to an assembly of rigid molecular dipoles, i.e. let there be no deformational polarisation. If the permanent moment of each dipole is μ then the energy u of such a dipole is given by

$$
u = -\boldsymbol{\mu} \cdot \boldsymbol{E}_{\mathrm{L}} = -\mu E_{\mathrm{L}} \cos \theta, \tag{2.24}
$$

where θ is the angle between the axis of the dipole and the direction of the local electric field $\boldsymbol{E}_{\mathrm{L}}$ (see fig. 2.4). If the electric field is the only orientational force acting on the molecules, the distribution of dipole orientations will be given by Boltzmann's law. Thus the number of dipoles pointing in the directions confined within the element of solid angle $\mathrm{d}\Omega$ will be

$$
A\mathrm{e}^{-u/kT}\mathrm{d}\Omega,
$$

where A is a constant which is determined by the total number of dipoles

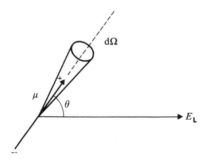

Fig. 2.4. Orientation of a molecular dipole in an applied electric field.

per unit volume. Hence the average molecular moment $\bar{m}$ in the direction of the electric field will be given by

$$\bar{m} = \frac{\int A e^{\mu E_L \cos \theta / kT} \mu \cos \theta \, d\Omega}{\int A e^{\mu E_L \cos \theta / kt} \, d\Omega}, \tag{2.25}$$

where the integration is taken over all possible directions.

Now the element of solid angle $d\Omega$, which includes all the directions in which the dipole lies at an angle in the range θ to $\theta + d\theta$ with respect to the direction of the applied field, is $d\Omega = 2\pi \sin \theta \, d\theta = 2\pi \, d(\cos \theta)$. Making this substitution and introducing the variable $x = \mu E_L / kT$, we obtain

$$\bar{m} = \frac{2\pi\mu \int_0^\pi e^{x \cos \theta} \cos \theta \, d(\cos \theta)}{2\pi \int_0^\pi e^{x \cos \theta} \, d(\cos \theta)}. \tag{2.26}$$

Integrating by parts:

$$\frac{\bar{m}}{\mu} = \frac{\left[\dfrac{\cos \theta}{x} e^{x \cos \theta} - \dfrac{1}{x^2} e^{x \cos \theta} \right]_0^\pi}{\left[\dfrac{1}{x} e^{x \cos \theta} \right]_0^\pi}$$

$$= \frac{\dfrac{1}{x}(e^x + e^{-x}) - \dfrac{1}{x^2}(e^x - e^{-x})}{\dfrac{1}{x}(e^x - e^{-x})}$$

$$= \frac{e^x + e^{-x}}{e^x - e^{-x}} - \frac{1}{x} = L(x), \tag{2.27}$$

where $L(x)$ is the Langevin function known from magnetic theory and may be expanded as a power series:

$$L(x) = \frac{x}{3} - \frac{x^3}{45} + \cdots \tag{2.28}$$

Fig. 2.5 shows a graph of the Langevin function. It is clear that there will be a *saturation* effect at very high fields. In practice such fields are

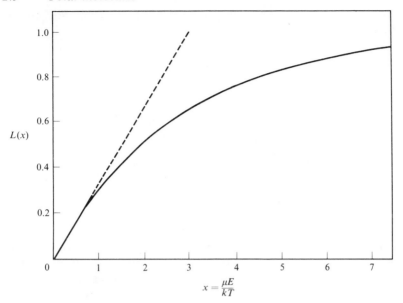

Fig. 2.5. Graph of the Langevin function, $L(x)$.

unapproachable and it is usually not necessary to take into account more than the first term of the Langevin function, so that

$$\bar{m} \approx \frac{\mu^2 E_L}{3kT}.$$ (2.29)

Consequently polarisation due to dipolar orientation is directly proportional to the local field strength and inversely proportional to temperature. We can think of the quantity $\mu^2/3kT$ as an orientational polarisability and simply add it to the normal deformational polarisability α_0 to give a total effective polarisability:

$$\alpha = \alpha_0 + \frac{\mu^2}{3kT}.$$ (2.30)

A more rigorous mathematical treatment that takes into account the possible anisotropy of the deformational polarisibility of molecules, i.e. that allows for the fact that an electric field can distort real molecules more easily in some directions than in others, requires that

$$\alpha_0 = \frac{\alpha_{11} + \alpha_{22} + \alpha_{33}}{3}$$ (2.31)

where α_{11}, α_{22} and α_{33} define the ellipsoid of the deformation-polarisability tensor.

The complete expression for the molar polarisation is then

$$P_{\text{M}} = \frac{\varepsilon - 1}{\varepsilon + 2} \frac{M_{\text{w}}}{\rho} = \frac{N_{\text{A}}}{3\varepsilon_0} \left(\frac{\alpha_{11} + \alpha_{22} + \alpha_{33}}{3} + \frac{\mu^2}{3kT} \right). \tag{2.32}$$

This formula provides a simple way of determining molecular dipole moments from experimental measurements. The method is to plot molar polarisation or effective polarisability against $1/T$, as in fig. 2.6; the slope of the straight line is then simply related to the square of the dipole moment. The method is most aptly applied to gases where one might expect the simple theory, which essentially neglects intermolecular effects, to be most accurate. The method may also be extended to dilute solutions of polar molecules in non-polar solvents. For this purpose simple additivity of solute and solvent polarisations is assumed, each component making a contribution proportional to its molar fraction f in the solution. The polarisation of the solution, experimentally observed through its dielectric constant, then complies with the following equation:

$$\frac{\varepsilon - 1}{\varepsilon + 2} \frac{M_{\text{w1}} f_1 + M_{\text{w2}} f_2}{\rho} = P_{\text{M1}} f_1 + P_{\text{M2}} f_2, \tag{2.33}$$

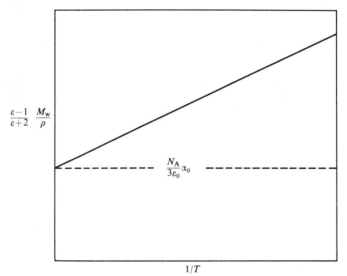

Fig. 2.6. Temperature dependence of molar polarisation.

where subscripts 1, 2 denote solute and solvent respectively. Knowing the molar polarisation of the pure solvent P_{M2}, the molar polarisation of the solute P_{M1} can be evaluated. In practice the latter varies with the concentration f_1, usually decreasing at high concentrations. This effect is attributed to association between solute molecules, and is eliminated by extrapolating the results to infinite dilution, where each solute molecule must be entirely surrounded by solvent molecules. This approach has been worked extensively, and a long list of molecular dipole moments has been compiled (Wesson, 1948). An illustrative sample of results is given in table 2.3. It is reasonable to suppose that various atomic groupings will have characteristic moments, and that when several groups are combined in one complicated molecule their moments will add vectorially. This principle has proved to be a very valuable test both for assessing theories about molecular configuration and for predicting values of dipole moments. Group moments depend considerably on the character of their bonding to the rest of the molecule in which they are present, however, and quantitative discrepancies are bound to occur. Thus the dipole moment of a substituent chlorine atom will be different in aliphatic and aromatic compounds, and two groups which are close to each other in a molecule may affect each other. Nevertheless, a useful assignment of group moments has been established and this enables one to make a good estimate of the dipole moment of most molecular configurations. For example, from the list of group moments given in

Table 2.3. *Molecular dipole moments*

Compound	Dipole moment ($\times 10^{-30}$ Cm)
H_2O	6.1
HF	6.4
HCl	3.6
HBr	2.6
CH_4	0
CCl_4	0
CO_2	0
NH_3	4.9
CH_3Cl	5.3
C_2H_5OH	5.7
$(C_2H_5)_2O$	3.8
C_6H_5Cl	5.7
C_6H_5Br	5.8
$C_6H_5NO_2$	14.0
$C_6H_5CH_3$	1.2

table 2.4, the predicted molecular moment for *p*-chloronitrobenzene is 8.3×10^{-30} C m, compared with the observed value 9.3×10^{-30} C m.

The directionality of group moments is usually clear from chemical evidence or by a study of the resultant dipole moment of a compound following the introduction of other polar substituents.

The situation is much more complicated in solids because the inter-molecular effects can no longer be ignored, i.e. the approximation $E_M = 0$ inherent in the simple formula for the local field (2.17) is not generally true. Consequently, although we can predict the molecular dipole moment from known group moments, it is not possible to

Table 2.4. *Group dipole moments*

Group	Aliphatic compounds Moment ($\times 10^{-30}$ Cm)	Angle (degrees)	Aromatic compounds Moment ($\times 10^{-30}$ Cm)	Angle (degrees)
–CH$_3$	0	0	1.3	0
–F	6.3		5.3	
–Cl	7.0		5.7	
–Br	6.7		5.7	
–I	6.3		5.7	
–OH	5.7	60	4.7	60
–NH$_2$	4.0	100	5.0	142
–COOH	5.7	74	5.3	74
–NO$_2$	12.3	0	14.0	0
–CN	13.4	0	14.7	0
–COOCH$_3$	6.0	70	6.0	70
–OCH$_3$	4.0	55	4.3	55

The angle denotes the direction of the moment with respect to the bond attaching the group to the rest of the molecule

calculate the molar polarisation and thereby the dielectric constant, without further elaboration of the dielectric model. In the case of a polymer there are further complications which arise from the flexibility of the long chains.

The weakness of the simple theory is highlighted if we solve the Debye version of the Clausius–Mosotti relation (2.32) for ε, neglecting the contribution of α_0:

$$\varepsilon = \frac{1 + [(2N_A \mu^2)/(9\varepsilon_0 kT)]}{1 - [(N_A \mu^2)/(9\varepsilon_0 kT)]}. \tag{2.34}$$

This predicts that as the temperature is decreased, ε becomes infinite at a temperature T_c given by

$$T_c = \frac{N_A \, \mu^2}{9k \, \varepsilon_o}, \tag{2.35}$$

and we may rewrite the relation in terms of this critical temperature:

$$\varepsilon - 1 = \frac{3T_c}{T - T_c}. \tag{2.36}$$

The implication is that lowering the temperature eventually allows the polarisation to increase to such an extent that the molecular dipoles may *spontaneously* align themselves parallel to each other, even in the absence of an applied field.

This is called the ferroelectric effect, by analogy with ferromagnetism; equation (2.36) is analogous to the Curie–Weiss law. A few crystalline materials do indeed display ferroelectric behaviour, but the phenomenon is rare. This means that the Debye equation fails for most materials in the regime of high local fields and it is necessary to improve the theoretical treatment to cope with this situation.

Onsager derived an improved formula by adopting a better model for the calculation of the local field at a molecule. His model consists of a spherical cavity which is excised in the dielectric material and which is just large enough to accommodate one molecule. The molecular dipole is supposed to be a point dipole μ at the centre of the sphere, radius a. Onsager then said that the local field operating on the dipole at the centre of the cavity could be resolved into two components, a cavity field G and a reaction field R:

$$E_L = G + R. \tag{2.37}$$

The cavity field is defined as that present at the centre of an empty cavity in the presence of the applied field E, i.e.

$$G = \frac{3\varepsilon}{2\varepsilon + 1} E. \tag{2.38}$$

The reaction field is defined as the extra field at the centre of the cavity due to the polarisation of the material external to the cavity caused by the presence of the point dipole at the centre of the cavity, i.e.

$$R = \frac{2(\varepsilon - 1)}{\varepsilon_0 (2\varepsilon + 1)a^3} \mu. \tag{2.39}$$

Using this local field one finally obtains the relation

$$\frac{(\varepsilon - n^2)(2\varepsilon + n^2)}{\varepsilon(n^2 + 2)^2} \frac{M_w}{\rho} = \frac{N_A \mu^2}{9\varepsilon_0 kT},\tag{2.40}$$

where the deformational polarisability of the molecules has been taken into account using the equation for molar refraction (2.23). This formula very successfully describes the behaviour of many fluids. It is an improvement on the Clausius–Mosotti relation in that it avoids artificially predicting a ferroelectric effect (often referred to as the ferroelectric catastrophe). Strong local forces have still been neglected, however, and it is therefore not surprising that it fails to explain the behaviour of associated liquids, like water, and many solids.

Kirkwood took a more rigorous statistical-mechanical approach in an attempt to incorporate the effect of local ordering. His theory is only valid for rigid dipoles, and it was left to Fröhlich to extend the treatment properly to a system of polarisable dipolar molecules. The work is well described in Fröhlich's classic text (1949). The final outcome was the following formula, which connects the macroscopic dielectric constant with molecular dipole moment:

$$\frac{(\varepsilon - n^2)(2\varepsilon + n^2)}{\varepsilon(n^2 + 2)^2} \frac{M_w}{\rho} = \frac{N_A g\mu^2}{9\varepsilon_0 kT}.\tag{2.41}$$

The parameter g is the correlation factor first introduced by Kirkwood to take account of local order:

$$g = 1 + z\, \overline{\cos \gamma},\tag{2.42}$$

where z is the number of nearest neighbours of a molecule in the system and γ is the angle of the reference molecule with respect to a nearest neighbour. The Fröhlich equation reduces to the Onsager equation where $g = 1$.

The usefulness of Fröhlich's formula (2.41) is mainly restricted by our ignorance of the correlation factor g, which necessarily depends on the shapes of molecules and the disposition of the permanent dipoles within them, the anisotropy of polarisability and the presence of charge distributions of higher orders of symmetry. The theory gives us a good general understanding of the behaviour of polar materials, but deviations from the simple Debye equation (2.32) can often only be discussed in qualitative terms.

Local ordering effects have long been recognised experimentally in measurements of dipole moments of polar solutes in non-polar solvents, where the value obtained on the basis of the simple model differs from the value obtained for the pure solute in the gas phase, even when the results are extrapolated to infinite dilution. This so-called solvent effect is due to the Onsager reaction field. If there is no strong local ordering, Onsager's formula (2.40) is valid and the apparent solution moment is related to the isolated molecule or gas moment by

$$\mu_{app} = \frac{(2\varepsilon_0 + 1)(n^2 + 2)}{3(2\varepsilon_0 + n^2)} \mu. \tag{2.43}$$

Usually, though, the solvent molecules do organise themselves in a special way around the polar species for the reason cited above. The effect depends principally on the anisotropy of the solute molecules, and many semi-empirical equations have been devised on this basis to explain the apparent change of dipole moments in solutions. An alternative approach has been to quote gas- and solution-determined dipole moments separately and to use the appropriate value in any particular situation.

In the case of polymers the principal ordering effect is an *intra*molecular one, as we shall see in the next section.

2.6   Dielectric constant of polymers

The foregoing theoretical development provides a general basis for our understanding and interpretation of the dielectric constants of polymers, which are our primary concern here. A convenient way of thinking about the dielectric behaviour of polymers is to consider not the complete long-chain molecules as the polarisable entities, but rather the component repeat units. For most polymeric materials, the degree of polymerisation is greater than a hundred, so that the effect of chain ends can be neglected for most purposes. The predominant constraint on each segment is due to its chemical linkage to the rest of the molecular chain on either side, and the strong intramolecular ordering, or correlation, of the segments along the chain can be formally taken into account through an intramolecular g-factor. We shall summarise and illustrate the results that have been obtained by this general approach, initially discussing a few overriding, simple principles, which give physical insight into dielectric properties of polymers.

2.6.1   Non-polar polymers

Let us consider an essentially non-polar polymer, e.g. polyethylene, CH_3–$(CH_2)_n$–CH_3. The density of solid polyethylene covers a range from 0.92 to 0.99 Mg m^{-3} depending on the extent of chain branching which determines its crystallinity. We may therefore test the validity of the Clausius–Mosotti relation. From published tables of bond polarisabilities, the Clausius–Mosotti relation for an assembly of $-CH_2-$ units becomes

$$\frac{\varepsilon-1}{\varepsilon+2}=K\rho, \quad \text{where } K=0.327, \tag{2.44}$$

i.e.

$$\varepsilon=2.276+2.01\,(\rho-920)\,10^{-3}.$$

Experimental measurements confirm the linear dependence of dielectric constant on density. Furthermore the agreement on the value of K (mean experimental value 0.326 compared with 0.327 theoretical) is very satisfactory.

2.6.2   Polar polymers

When permanent dipole moments are present in a polymeric chain we can distinguish two different situations: one where the whole polymer backbone, together with its polar groups, is rigidly fixed in a single conformation, and one where the backbone is flexible and pendant side groups can rotate freely. The former is more characteristic of the crystalline state than of amorphous or fluid states, although occasionally a particular configuration may be especially favoured and survive even in non-crystalline environments.

If a polymer is held in a fixed conformation, the resulting moment of a complete molecule will depend very much on whether the moments of the individual segments reinforce or compensate each other. In extended configurations of polytetrafluoroethylene, the high dipole moments of alternate $-CF_2-$ groups balance each other exactly (fig. 2.7(*a*)) and for this reason the dielectric constant of this polymer is low, like that of a non-polar material. However, some defects are always present in the conformations of the molecules (helical conformations, which also give dipole balancing, are typical of the crystalline phase of this polymer), and these are responsible for the small dipolar orientation effects that can just be detected. In contrast, the C–Cl dipole moments in poly(vinyl chloride) are additive in the preferred planar zigzag conformation (fig. 2.7(*b*)), so that the dielectric constant of this polymer is high.

A particularly striking additive effect is observed in the case of certain synthetic polypeptides, e.g. poly(γ-benzyl-L-glutamate) which has a formula $(-CO \cdot CHR \cdot NH-)_n$, $R=-CH_2 \cdot CH_2 \cdot CO \cdot O \cdot CH_2 \cdot C_6H_5$. In solution this substance readily forms an α-helix, maintained by hydrogen bonding, with an axial dipole moment per repeat unit of about

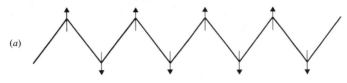

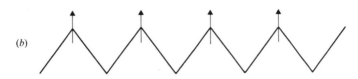

Fig. 2.7. Schematic arrangement of polar groups in a polymer chain: (a) balanced dipoles, (b) additive dipoles.

4.14×10^{-30} Cm. Since a typical molecular weight is about 500 000, the total axial dipole moment for a molecule is about $10\,000 \times 10^{-30}$ C m.

2.6.3   Mean-square moment

Most usually, polymer molecules will not be in a single, fixed conformation, and the experimentally observable quantity, the mean-square dipole moment, is an average over many different conformations. At any given instant the resultant moment M of a whole molecule will be given by the vector sum of all the segment moments, m_k:

$$M = \sum_{k=1}^{n} m_k. \tag{2.45}$$

The mean-square moment for an assembly of such molecules will be defined theoretically by

$$\overline{M^2} = \overline{\sum_{i=1}^{n} m_i \cdot \sum_{j=1}^{n} m_j}$$

$$= \sum_{i=1}^{n} \sum_{j=1}^{n} \overline{m_i \cdot m_j}$$

$$= m^2 \left(n + \sum_{\substack{i=1 \\ (i \neq j)}}^{n} \sum_{j=1}^{n} \overline{\cos \theta_{ij}} \right), \tag{2.46}$$

where $\overline{\cos \theta_{ij}}$ is the cosine of the angle θ_{ij} made between the dipole of repeat unit i and that of unit j averaged over all polymer molecules. In a real situation this depends principally on the constraints of chemical bonding, steric hindrance between neighbouring parts of the chain and upon dipole–dipole interaction along the chain, as well as interactions with neighbouring molecules. From equation (2.46) we may obtain an expression for the effective mean-square moment per repeat unit of the polymeric chains, and thereby define a segmental correlation factor g_r for the polymer:

$$\frac{\overline{M^2}}{n} = m^2 \left(1 + \frac{1}{n} \sum_{\substack{i=1 \\ (i \neq j)}}^{n} \sum_{j=1}^{n} \overline{\cos \theta_{ij}} \right) \tag{2.47}$$

$$= g_r m^2.$$

Intramolecular forces must dominate the correlation, and this notion provides the basis for a theoretical discussion of g_r-values. Intermolecular forces may be taken into account as a secondary perturbation.

An important case concerns the arrangement of segmental dipoles, aligned perpendicularly with respect to the chain contour, along a polymeric molecule which has a backbone of carbon atoms. Part of the dipolar correlation is fixed, of course, by the tetrahedral nature of the carbon valence, but part depends on the possible rotation about the C–C bonds of the chain. For completely free rotation it may be shown that $g_r = 11/12$. Only rarely is this situation approached in real molecules, and in the n-paraffins, for example, the *trans*-isomer is about 3 kJ mol^{-1} lower in energy than the *gauche*-isomer, so that at room temperature the *trans*-isomer is appreciably favoured, and the g_r-value for the CH$_2$ dipoles is reduced.

The series of polyethers of general formula $[-(CH_2)_x-O-]_n$, has been treated theoretically in detail (Read, 1965) and illustrates the principles of segmental dipole correlation very well. Confining ourselves here to a qualitative discussion, we first note that the main dipole moment, essentially the same in all of these molecules, is that of the oxygen atom in the chain ($\mu = 3.8 \times 10^{-30}$ Cm). In the simplest member of the series, poly(methylene oxide), the adjacent dipoles along the chain are very close to one another and the dipole–dipole repulsion is strong. For this reason a helical, *gauche*-conformation, which gives a low $\overline{M^2}/n$ value by dipole balancing, is preferred energetically. The low dielectric constant ($\varepsilon \approx 3.5$) is observed to increase with temperature and this may be explained by an increase in the higher energy *trans*-conformations, which produce parallel alignment of adjacent dipoles, as the thermal energy increases.

In the next member of the series, poly(ethylene oxide), the dipoles are further apart. An intermediate $\overline{M^2}/n$ value with negligible temperature dependence is observed, suggesting that the *trans-* and *gauche*-isomers have approximately equal energies. At room temperature $\varepsilon \approx 4.5$.

By the time the third member of the series, poly(trimethylene oxide), is reached, dipole–dipole interaction along the chain is very weak indeed, and the *trans*-conformation is slightly more favoured energetically (by about $4\,\text{kJ}\,\text{mol}^{-1}$). Since the *trans*-conformation involves parallel alignment of the oxygen dipoles, the observed decrease of dielectric constant with temperature is expected.

Finally, in poly(tetramethylene oxide), the *trans*-conformation is again the most favoured, but here the *trans*-conformation incurs anti-parallel alignment of the oxygen dipoles. The dielectric constant therefore increases with temperature.

Strictly speaking the foregoing discussion of segmental correlation has assumed that the polymers are isotactic. When the chemical repeat units along the polymeric chains are not all stereo-identical and syndio-tactic or even atactic isomers are present, the simple molecular concept of a segmental correlation factor no longer applies. A more elaborate treatment is then required (Volkenstein, 1963).

2.7   Further reading

The original monographs by Debye (1929) and Fröhlich (1949) still provide an excellent introduction to molecular dielectric theory. Smith (1955) gives detailed information about measurements of dipole moments and data on polymers are reviewed by McCrum, Read and Williams (1967).

3   Dielectric relaxation

3.1   General theory

Orientation of molecular dipoles is a relatively slow process if comparison is made with electronic transitions or molecular vibrations which have frequencies generally above 10^{12} Hz. Furthermore, it does not consist of a uniform switch in the arrangement of all molecules; it is more in the nature of a slight adjustment of their average orientations in the face of continued thermal agitation. Only when sufficient time is allowed after the application of an electric field for the orientation to attain equilibrium will the maximum polarisation, corresponding to the highest observable dielectric constant, be realised in a material. If ample time *is* allowed, then the observed dielectric constant is called the static dielectric constant ε_s. On the other hand, if the polarisation is measured immediately after the field is applied, allowing no time for dipole orientation to take place, then the observed instantaneous dielectric constant, denoted ε_∞, will be low and due to deformational effects alone. Somewhere in between these extremes of timescale there must therefore be a dispersion from a high to a low dielectric constant.

Let us start our examination of this rate effect by considering the application of an alternating electric field E, amplitude E_0 and angular frequency ω, across a dielectric material:

$$E = E_0 \cos \omega t. \tag{3.1}$$

This will produce polarisation which alternates in direction, and, if the frequency is high enough, the orientation of any dipoles which are present will inevitably lag behind the applied field. Mathematically, we can express this as a phase lag δ in the electric displacement:

$$D = D_0 \cos (\omega t - \delta), \tag{3.2}$$

which may be written

$$D = D_1 \cos \omega t + D_2 \sin \omega t, \tag{3.3}$$

where

$$D_1 = D_0 \cos \delta \quad \text{and} \quad D_2 = D_0 \sin \delta.$$

This leads us to define two dielectric constants

$$\varepsilon' = \frac{D_1}{\varepsilon_0 E_0} \quad \text{and} \quad \varepsilon'' = \frac{D_2}{\varepsilon_0 E_0}, \tag{3.4}$$

linked by the relation

$$\frac{\varepsilon''}{\varepsilon'} = \tan \delta. \tag{3.5}$$

It is convenient to combine these two quantities into a complex dielectric constant or relative permittivity:

$$\varepsilon^* = \varepsilon' - i \varepsilon''. \tag{3.6}$$

The meaning of the real and imaginary parts may be readily appreciated by considering the material in a capacitor (capacitance C_0 when empty), as shown in fig. 3.1. The current I which flows in the external circuit after application of an alternating voltage given by the real part of $V = V_0 e^{i\omega t}$, may be calculated as follows:

$$I = \varepsilon^* C_0 \frac{dV}{dt}$$
$$= i\omega \varepsilon^* C_0 V$$
$$= \omega C_0 (\varepsilon'' + i\varepsilon') V. \tag{3.7}$$

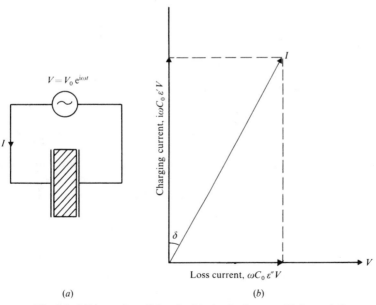

(a)   (b)

Fig. 3.1. AC losses in a dielectric: (a) circuit diagram, (b) Argand diagram of complex current–voltage relationship.

This implies that we have a capacitive component of the current,

$$I_C = i\omega C_0 \varepsilon' V, \tag{3.8}$$

which leads the voltage by $90°$, and a resistive component,

$$I_R = \omega C_0 \varepsilon'' V, \tag{3.9}$$

which is in phase with the voltage. Work can only be done by the latter component, and the physical meaning of the useful quantity tan δ, previously defined by equation (3.5), becomes apparent:

$$\tan \delta = \frac{\varepsilon''}{\varepsilon'} = \frac{\text{energy dissipated per cycle}}{\text{energy stored per cycle}}. \tag{3.10}$$

ε'' is called the dielectric loss factor and tan δ is usually called the dielectric loss tangent or dissipation factor.

ε' and ε'' are experimentally observable quantities which may be used to characterise the dielectric dispersion over a range of frequencies. In order to be able to interpret any such dispersion behaviour, it is first necessary to forge a link between these macroscopic, observable quantities and molecular properties, by postulating a reasonable model which describes the way the molecules respond to the applied field.

The basic theory of dielectric relaxation behaviour, pioneered by Debye, begins with a macroscopic treatment of frequency dependence. This treatment rests on two essential premises: exponential approach to equilibrium and the applicability of the superposition principle. In outline, the argument is as follows.

Consider the application at time $t=0$ of a steady field E_0 across a dielectric. The resulting electric displacement $D(t)$ at subsequent times t will then comply with the equation:

$$D(t) = \varepsilon_0[\varepsilon_\infty + (\varepsilon_s - \varepsilon_\infty)\Psi(t)]E_0. \tag{3.11}$$

The first term on the right-hand side, $\varepsilon_0\varepsilon_\infty E_0$, represents the instantaneous response of the material to the field. The second term, $\varepsilon_0 (\varepsilon_s - \varepsilon_\infty)\Psi(t)E_0$, represents the slower contribution from polarisation of dipoles, with the factor $\Psi(t)$ describing the time development of the underlying orientation process. By this definition $\Psi(0)=0$ and $\Psi(\infty)=1$. We further assume that the rates at which dipolar polarisation $^D P(t)$ progresses towards its equilibrium value $^D P(\infty) = \varepsilon_0(\varepsilon_s - \varepsilon_\infty)E_0$ is proportional to its degree of departure from equilibrium, i.e.

$$^D\dot{P}(t) = -\frac{^D P(\infty) - {}^D P(t)}{\tau}. \tag{3.12}$$

Here τ is a characteristic time constant, usually called the dielectric relaxation time. To conform with analogous theories of visco-elastic behaviour we should really use the term dielectric *retardation* time, because it refers to a gradual change in a strain (the polarisation or resulting electric displacement) following an abrupt change in stress (the applied field). Dielectric *relaxation* time is still most commonly used, however, in spite of this inconsistency. By integration of equation (3.12):

$$^D\!P(t) = {}^D\!P(\infty)\,(1 - e^{-t/\tau}), \tag{3.13}$$

so that

$$\Psi(t) = 1 - e^{-t/\tau}. \tag{3.14}$$

If polarisation of dipoles is a linear function of applied field, then a higher field $E_0 + E_1$ applied at time $t = 0$ will produce a proportional increase in electric displacement at all subsequent times:

$$D(t) = \varepsilon_0 [\varepsilon_\infty + (\varepsilon_s - \varepsilon_\infty)\Psi(t)](E_0 + E_1) \tag{3.15}$$

According to Boltzmann's superposition principle for such linear systems, if the extra field is added at time t_1, the total displacement at times $t > t_1$ will be

$$D(t) = \varepsilon_0 [\varepsilon_\infty + (\varepsilon_s - \varepsilon_\infty)\Psi(t)]E_0 + \tag{3.16}$$
$$\varepsilon_0 [\varepsilon_\infty + (\varepsilon_s - \varepsilon_\infty)\Psi(t - t_1)]E_1,$$

i.e. the simple adduct of the two electric displacements at their respective elapsed times. In general, for a series of field increments

$$D(t) = \sum_{t_i = -\infty}^{t_i = t} \varepsilon_0 [\varepsilon_\infty + (\varepsilon_s - \varepsilon_\infty)\Psi(t - t_i)]E_i, \tag{3.17}$$

whence for a continuously varying field

$$D(t) = \varepsilon_0 \varepsilon_\infty E(t) + \int_{-\infty}^{t} \varepsilon_0(\varepsilon_s - \varepsilon_\infty)\Psi(t - s)\frac{dE}{ds}ds. \tag{3.18}$$

Equation (3.18) expresses the value of the electric displacement actually present in the material at time t in terms of the entire past history of subjection to applied field. Integrating by parts:

$$D(t) = \varepsilon_0 \varepsilon_\infty E(t) + \varepsilon_0(\varepsilon_s - \varepsilon_\infty)\int_{-\infty}^{t} \dot{\Psi}(t - s)E(s)ds. \tag{3.19}$$

From equation (3.14) the function $\dot{\Psi}(t)$, called the *dielectric decay function*, is given by

$$\dot{\Psi}(t) = \frac{1}{\tau}e^{-t/\tau}. \tag{3.20}$$

Now differentiation of equation (3.19) with respect to t gives

$$\frac{\mathrm{d}D(t)}{\mathrm{d}t} = \varepsilon_0\varepsilon_\infty\frac{\mathrm{d}E(t)}{\mathrm{d}t} + \varepsilon_0(\varepsilon_s - \varepsilon_\infty)\frac{\mathrm{d}}{\mathrm{d}t}\int_{-\infty}^{t}\dot{\Psi}(t-s)E(s)\,\mathrm{d}s, \tag{3.21}$$

and from equation (3.20) it follows that

$$\frac{\mathrm{d}}{\mathrm{d}t}\int_{-\infty}^{t}\dot{\Psi}(t-s)E(s)\,\mathrm{d}s = \dot{\Psi}(0)E(0) - \int_{-\infty}^{t}\frac{1}{\tau}\dot{\Psi}(t-s)E(s)\,\mathrm{d}s. \tag{3.22}$$

Since, also from equation (3.20), $\dot{\Psi}(0) = 1/\tau$, combining equations (3.19), (3.21) and (3.22) finally gives the following differential equation for the electric displacement in the material:

$$\tau\frac{\mathrm{d}D(t)}{\mathrm{d}t} + D(t) = \tau\varepsilon_0\varepsilon_\infty\frac{\mathrm{d}E(t)}{\mathrm{d}t} + \varepsilon_0\varepsilon_s E(t). \tag{3.23}$$

For the particular case of an alternating field represented by the real part of $E(t) = E_0\mathrm{e}^{i\omega t}$, which produces an alternating displacement $D(t) = D_0\mathrm{e}^{i\omega(t-\delta)}$, the solution to equation (3.23) may be written

$$\varepsilon^* = \frac{D(t)}{\varepsilon_0 E(t)} = \varepsilon_\infty + \frac{\varepsilon_s - \varepsilon_\infty}{1 + i\omega\tau}, \tag{3.24}$$

which is called the Debye dispersion equation. Equating real and imaginary parts of the two sides,

$$\varepsilon'(\omega) = \varepsilon_\infty + \frac{\varepsilon_s - \varepsilon_\infty}{1 + \omega^2\tau^2}, \tag{3.25}$$

$$\varepsilon''(\omega) = \frac{\varepsilon_s - \varepsilon_\infty}{1 + \omega^2\tau^2}\omega\tau. \tag{3.26}$$

The graphs of ε' and ε'' against frequency of the applied field (logarithmic scale) through the dispersion region are shown in fig. 3.2. The dielectric loss peak has a half-height width of 1.14 decades. The maximum loss value occurs when $\omega\tau = 1$, corresponding to a critical frequency $\omega_{max} = 1/\tau$, and location of this peak provides the easiest way of obtaining the relaxation time from experimental results. The difference in dielectric constant measured at low and high frequencies is called the *strength* of the relaxation and it is related to the area under the absorption curve:

$$\Delta\varepsilon = \varepsilon_s - \varepsilon_\infty = \frac{2}{\pi}\int_{-\infty}^{+\infty}\varepsilon''(\omega)\mathrm{d}(\ln\omega). \tag{3.27}$$

(This is a particular case of the Kramers–Krönig relation which expresses the general interdependence of quantities like ε'' and ε' for any linear system.)

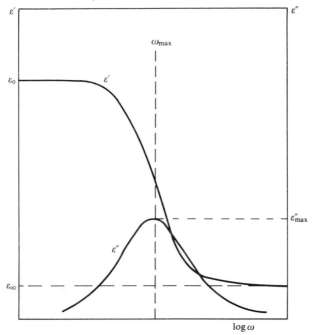

Fig. 3.2. Debye dielectric dispersion curves.

The dependence of ε'' on ε' may be used to test how well the Debye model fits a real case. If we eliminate the parameter $\omega\tau$ between equations (3.25) and (3.26), we obtain

$$\left(\varepsilon' - \frac{\varepsilon_s + \varepsilon_\infty}{2}\right)^2 + \varepsilon''^2 = \left(\frac{\varepsilon_s - \varepsilon_\infty}{2}\right)^2. \tag{3.28}$$

This is the equation of a circle, centre $[(\varepsilon_s + \varepsilon_\infty)/2, 0]$, radius $(\varepsilon_s - \varepsilon_\infty)/2$, so that a plot of ε'' against ε' should give a semicircle, as shown in fig. 3.3. Experimental results for many polar liquids give excellent agreement with this theoretical curve, their relaxation times being of the order of 10^{-11} s.

Relaxations observed in polymers, however, show broader dispersion curves and lower loss maxima than those predicted by the Debye model, and the $\varepsilon'' - \varepsilon'$ curve falls inside the semicircle. This led Cole and Cole (1941) to suggest the following semi-empirical equation for dielectric relaxations in polymers:

$$\varepsilon^* = \varepsilon_\infty + \frac{\varepsilon_0 - \varepsilon_\infty}{1 + (i\omega\tau)^\alpha}, \tag{3.29}$$

where α is a parameter, $0 < \alpha \leq 1$.

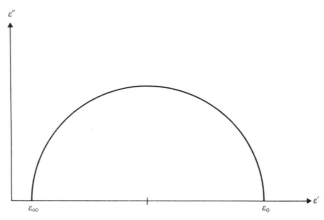

Fig. 3.3. Semicircular Cole–Cole plot.

Equation (3.29) better describes a broad dispersion, and gives a ε''–ε' plot where the centre of the semicircle is depressed below the abscissa. It corresponds to a superposition of a group of Debye-like relaxation processes with a range of relaxation times that are symmetrically distributed about τ. Such a spread of relaxation times was seen as a very likely explanation of the broadness of the relaxation in a system of entangled, long-chain molecules where the forces restraining the orientation of segmental dipoles might be expected to cover a very wide range. Although the exact form of the distribution of relaxation times in the Cole–Cole equation is complicated and is not based on any special model, the parameter α is a convenient one for specifying the broadness of experimental relaxation peaks, and it has been used extensively for this purpose. Davidson and Cole (1950) improved the fit with experiment by using a slightly different semi-empirical equation:

$$\varepsilon^* = \varepsilon_\infty + \frac{\varepsilon_s - \varepsilon_\infty}{(1 + i\omega\tau)^\beta}, \tag{3.30}$$

where β is a parameter, $0 < \beta \le 1$.

Equation (3.30) corresponds to a skewed distribution of relaxation times about τ, but it again has no particular theoretical foundation apart from the improved agreement with experiment for certain materials.

More recently, it has been suggested that a departure from the assumed exponential form of the approach to equilibrium may be responsible for the breadth of dipolar relaxation in polymers. A decay function of the form

$$\Psi(t) = e^{-(t/\tau)^\gamma}, \tag{3.31}$$

with $0 < \gamma \leq 1$, gives very good agreement in the case of many polymers, but the physical explanation for this kind of decay function has not yet been established (Cook, Watts and Williams, 1970).

3.2   Thermal activation of dipolar relaxation

In order to progress further in the interpretation of dipolar relaxation behaviour we must develop a molecular model in still more detail. Many different models, depending on the type of material concerned, have been proposed and these have led to a large number of theoretical treatments. We shall confine ourselves to two theories, which are particularly relevant to polymers, and provide a suitable basis for general

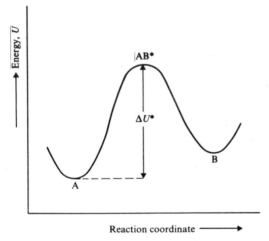

Fig. 3.4. Potential-energy diagram for a thermally activated process.

discussion of the main features encountered. First, in this section, we will consider the theory which has been central to the understanding of the temperature dependence of nearly all reaction rate processes: thermal activation over a potential-energy barrier.

The model (fig. 3.4) simply consists of two molecular states, A and B, separated by a potential-energy barrier of height ΔU^*, and the reaction whose rate is calculated is the change A⟶B. A and B may be taken to represent two orientations of a dipolar group about a valence bond linking the group to the rest of the molecule, the *reaction coordinate* being the angle of rotation about the bond. The reaction will follow the usual monomolecular rate equation:

$$-\frac{\mathrm{d}c_A}{\mathrm{d}t} = k_A\, c_A, \qquad\qquad (3.32)$$

where c_A is the concentration of dipoles in state A and k_A is the rate constant. It is clear that molecules must acquire an extra amount of energy ΔU^* in order to surmount the energy barrier prior to passing from state A to state B. Now it is well known from Boltzmann statistics that the probability that a given molecule possesses energy in excess of an amount ΔU^* is proportional to $e^{-\Delta U^*/kT}$, so that we may expect the rate constant k_A to be given by

$$k_A = Ae^{-\Delta U^*/kT} \tag{3.33}$$

where A is a constant or a function which only varies slowly with temperature, and ΔU^* is called the activation energy. The relaxation time τ may be roughly identified with $1/k_A$, so that we shall have

$$\ln\tau = \frac{\Delta U^*}{kT} + \text{constant}. \tag{3.34}$$

This equation, an example of the well-known Arrhenius law, means that a plot of $\ln\tau$ against $1/T$ should give a straight line whose slope is directly related to the activation energy.

   In order to calculate an absolute reaction rate we may turn to transition-state theory (Glasstone, Laidler and Eyring, 1941). In this theory the intermediate stage in the reaction, corresponding to the peak in the potential-energy curve and called the transition state AB*, is treated as a pseudo-stable state whose equilibrium concentration can be calculated in terms of statistical mechanical partition functions. A partition function summarises the distribution of molecules over all their possible energy states and it determines the free energy of a system. It may be factorised with respect to each independent degree of freedom (Ubbelohde, 1952). The partition functions are somewhat arbitrary for the transition state, because there is one special degree of freedom – the *vibration* over the potential hill or pass along the reaction coordinate. It is supposed that the frequency v of this slow (negative force constant) vibration gives the velocity of the change from the transition state to the final state. On this basis,

$$k_A = K^*v, \tag{3.35}$$

where K^* is the chemical equilibrium constant for the transition state:

$$K^* = \frac{c_{AB^*}}{c_A}, \tag{3.36}$$

where c_{AB^*} is the concentration of dipoles in the transition state. Applying the usual principle of minimisation of free energy to the equilibrium, we shall also have

$$K^* = e^{-\Delta G^*/RT}, \tag{3.37}$$

where ΔG^* is the change in molar free energy in going to the transition state. Eyring suggests, however, that it is appropriate to separate out the partition function $kT/h\nu$ for the degree of freedom belonging to the vibration in the reaction coordinate, because this degree of freedom is only accessible to activated molecules, so that

$$K^* = \frac{kT}{h\nu} e^{-\Delta G^*/RT}. \qquad (3.38)$$

Hence by substituting equation (3.38) in equation (3.35)

$$k_A = \frac{kT}{h} e^{-\Delta G^*/RT}, \qquad (3.39)$$

and therefore

$$\tau = \frac{1}{k_A}$$

$$= \frac{h}{kT} e^{\Delta G^*/RT} = \frac{h}{kT} e^{-\Delta S^*/R} e^{\Delta H^*/RT}, \qquad (3.40)$$

where ΔS^* is the molar entropy of activation and ΔH^* is the molar enthalpy of activation. We see that the pre-exponential factor is dependent on temperature, although the temperature dependence of the relaxation time will be dominated by the exponential factor. Neglecting the entropy factor, we calculate that, at room temperature,

$$\tau \approx 10^{-12} e^{\Delta H^*/RT}. \qquad (3.41)$$

3.3 Cooperative dipolar relaxation in polymers

A major feature of the behaviour of amorphous or partially crystalline polymeric materials is the glass transition. At low temperatures most plastics become hard and brittle, whereas at high temperatures they are rubbery or leathery and have great flexibility and toughness. The change from one form to the other is found to take place over a comparatively narrow range of temperature. It can be demonstrated by measuring the force needed to push a needle into a polymer at a given rate, or observed more qualitatively as a change in the ease of folding a sheet of the material. The change is not a first-order transition – no latent heat is given out or taken in and there are no discontinuities in density and other properties – but sharp changes in temperature dependence of various properties are seen, suggesting a second-order transition. The most usual way of determining the glass transition temperature T_g is to follow the volume contraction of a sample in a dilatometer and to note the temperature at which the volume–temperature graph changes in slope (fig. 3.5). The T_g value found in this way depends on the rate of cooling, e.g. polystyrene gives the following results:

$T_g = 105\ °C$ at 1 deg min^{-1},
$T_g = 100\ °C$ at 1 deg day^{-1}.

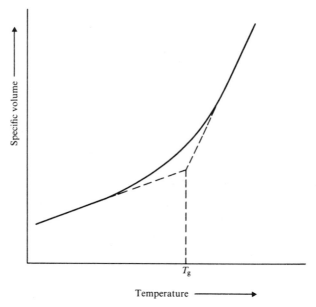

Fig. 3.5. Volumetric determination of the glass transition.

We may attribute T_g to a major change in the segmental mobility of the polymer chains. Above T_g there is sufficient mobility, a sort of micro-Brownian motion, which enables large-scale reorganisation of the chains to occur in response to an applied stress (e.g. a change of temperature), whereas below T_g the chains are *frozen* in position. In other words, an observed T_g is that temperature at which the time constant for a molecular rearrangement process becomes comparable with the timescale of the experiment used to measure it. As may be expected, at the onset of the molecular mobility above T_g, permanent dipoles attached rigidly to the polymer backbone become free to orient in an electric field, and the glass transition is accompanied by a major dielectric dispersion.

The temperature dependence of the dielectric relaxation time of the molecular process associated with T_g does not conform with the simple Arrhenius law. The plot of $\ln \tau$ against $1/T$ is curved, as if the activation energy were increasing towards lower temperatures. This effect is seen in other properties too, and the general inference is that a large-scale rearrangement of a long-chain molecule involves a cooperative mechanism, i.e. the movement of one molecule is not independent of its neighbours. One way of looking at this is in terms of *free volume*: for a molecular segment to move or twist there must be a vacant site for it to move into. The availability of vacant sites may be expressed as an average free volume per molecular segment v_f, defined by

$$v_f = v - v_0, \tag{3.42}$$

where v is the actual volume occupied by a segment, and v_0 is the close-packed-sphere volume, approximately equal to the volume per segment at $0 \,°K$. Free volume increases with temperature, as evidenced by the relatively high coefficient of expansion of a rubber. This would be predicted on the basis that an extra amount of energy is associated with a hole, so that the probability of a hole existing in a system will increase with temperature in accord with the Boltzmann distribution law. At low temperatures, where free volume is scarce, the occurrence of a hole of the right size at the right place may become the controlling factor in molecular movement. By considering the probability of holes joining together to provide the *critical free volume* v_f^* which is needed for a segmental jump to take place, it may be shown (Bueche, 1962) that the dependence of the rate r of segmental motion on free volume takes the form

$$r \propto \exp(-v_f^*/v_f). \tag{3.43}$$

(This explains the success of the Doolittle empirical equation for the viscosity of low molecular weight hydrocarbon liquids:

$$\eta \propto \exp(1/f), \tag{3.44}$$

where f is the fractional free volume, v_f/v.) Now consider the ratio of the rates of segmental movement for a polymer at two different temperatures, T_1 and T_2, where the segmental free volumes are v_{f_1}, and v_{f_2}, respectively:

$$\ln \frac{r_2}{r_1} = v_f^* \left(\frac{1}{v_{f_1}} - \frac{1}{v_{f_2}} \right). \tag{3.45}$$

If we assume that the excess rate of expansion of the rubber over that of the glass is entirely attributable to an increasing free volume, we may write

$$v_{f_2} = v_{f_1} + \alpha v_1 (T_2 - T_1),$$ (3.46)

where α is the difference between the cubical expansion coefficients above and below the glass transition temperature, and v_1 is the actual segmental volume at temperature T_1. Then substituting for v_{f_2} in equation (3.45),

$$\ln \frac{r_2}{r_1} = \frac{(v_f^*/v_{f_1})(T_2 - T_1)}{(v_{f_1}/\alpha v_1) + T_2 - T_1}.$$ (3.47)

This equation has the same form as the well-known WLF equation (Williams, Landel and Ferry, 1955) which correlates the mechanical behaviour of all polymers near their T_g, provided we set $T_1 = T_g$ (T_g measured by the same method for each polymer). From experimental results one finds that

$$\frac{v^*}{v_{f_g}} \approx 40 \quad \text{and} \quad \frac{v_{f_g}}{v_1 \alpha} \approx 52,$$ (3.48)

which imply that $v^* \approx$ (size of a molecular segment). The WLF equation fits dielectric data if we set

$$\frac{r_2}{r_1} = \frac{\tau_1}{\tau_2},$$ (3.49)

i.e. if we take the dipolar relaxation time as a measure of segmental mobility. The temperature dependence of relaxation may then be written in the form

$$\log a_T = \log \frac{\tau_T}{\tau_{T_g}} = -\frac{C_1 (T - T_g)}{C_2 + T - T_g},$$ (3.50)

where C_1, C_2 are universal constants.

3.4   Dielectric relaxation in solid polymers

Several distinct dielectric relaxation processes are usually present in a solid polymeric material. This multiplicity is seen most easily in a scan of dielectric loss at constant frequency as a function of temperature (fig. 3.6). As the temperature is raised, molecular mobilities of various types become successively energised and available for dipolar orientation. By convention the dielectric relaxation processes are labelled α, β ... and so on, beginning at the high-temperature end. The same relaxation processes are generally responsible for dispersions in mechanical properties too, although a particular molecular rearrangement process may produce a stronger dielectric than mechanical effect, or vice versa.

Some polymers are wholly amorphous and there is only one phase present in the solid material. In such cases there is always a high-temperature α-relaxation associated with the micro-Brownian motion of the whole chains and, in addition, at least one low-temperature (β, γ etc.) subsidiary relaxation. The relative strength of the α- and β-dielectric relaxations depends on how much orientation of the dipolar groups can occur through the limited mobility allowed by the β-process before the

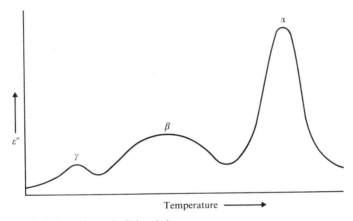

Fig. 3.6. A schematic dielectric loss curve.

more difficult but more extensive mobility of the α-process comes into play: there is a partitioning of the total dipolar alignment amongst the molecular rearrangement processes.

Detailed examination of the relaxations requires isothermal scans of dielectric constant and loss as a function of frequency f so that effective dipole movements and activation energies of relaxation times may be obtained. A typical pair of plots of ε' and ε'' values against $\log f$ is shown in fig. 3.7. Graphs of dielectric data of this kind are sometimes called, rather loosely, dielectric spectra. From a series of such plots the relaxation times can be obtained for the individual relaxation processes as a function of temperature.

The high-frequency β, γ . . . subsidiary peaks in amorphous polymers are characteristically very broad with a half-height width of several decades (compared with 1.14 decades for a single Debye relaxation process), although a good, linear Arrhenius plot is usually obtained

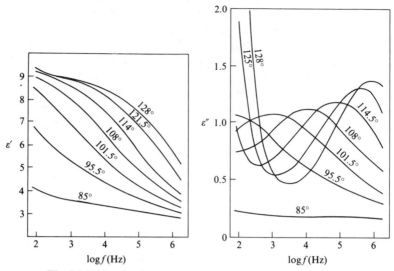

Fig. 3.7. Dielectric relaxation curves for poly(vinyl chloride) in the α-relaxation region. From Ishida (1960), by courtesy of Dr Dietrich Steinkopff Verlag.

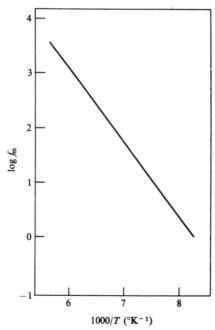

Fig. 3.8. Temperature dependence of the β-relaxation process in polyepichlorhydrin. Reprinted from Blythe and Jeffs (1969), p. 141, by courtesy of Marcel Dekker, Inc.

suggesting a non-cooperative mechanism. Fig. 3.8 shows the Arrhenius plot for the β-relaxation of polyepichlorhydrin,

$$CH_2Cl$$
$$|$$
$$(-CH-CH_2-O-)_n$$

where the β-process is attributable to rotation of the polar $-CH_2Cl$ group about its C–C linkage to the main polymer chain (Blythe and Jeffs, 1969). Relatively small activation energies are found for β-type processes and values for representative polymers are listed in table 3.1. The mechanism of a β-type process may be one of several different types depending on the nature of the dipole group concerned and its position

Table 3.1. *Activation energies for β-relaxa-*
tion in solid polymers

Polymer	$\Delta H^*(kJ\ mol^{-1})$
Poly(methyl methacrylate)	84
Polyepichlorhydrin	35
Poly(vinyl chloride)	63
Poly(vinyl acetate)	42
Poly(4-chlorocyclohexyl methacrylate)	48

on the polymer chain. Among the most important mechanisms are the following:

(*a*) *Rotation of a side group about a C–C bond.* This is probably the simplest type conceptually. It may be a small group, e.g. $-CH_2Cl$, or a more complicated side chain, e.g. $-CO \cdot OC_2H_5$.

(*b*) *Conformational flip of a cyclic unit.* The best-known examples involve the cyclohexyl side group. The transition from one chair-form to another alters the orientation of a polar substituent (fig. 3.9 (*a*)).

(*c*) *Local motion of a segment of the main chain.* Some limited, localised movement of the main chain is a common feature. This has to be the explanation of the β-process in poly(vinyl chloride), for example, where the dipolar group is directly attached to the main chain and cannot move independently of the polymer backbone. This kind of movement also arises when there are runs of four or more CH_2 units (Willbourn, 1958) and can be explained as a *crankshaft* rotation (Schatzki, 1962) about two collinear C–C bonds. The smallest segment of a $(CH_2)_n$ chain which can allow such rotation, whilst not involving the rest of the chain, is a $(CH_2)_4$ group (fig. 3.9(*b*)).

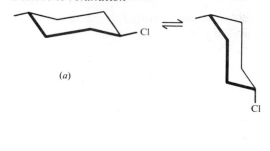

(a)

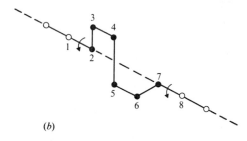

(b)

Fig. 3.9. Diagrams of molecular relaxation mechanisms: (a) conformational 'flip' of chlorohexane, (b) 'crankshaft' rotation in polyethylene.

The $\alpha(T_g)$-relaxation peak of an amorphous polymer is typically much narrower than a β-peak, although still considerably broader than that for a simple Debye process. Also the temperature dependence of the α-process is generally much steeper than that of a β-process, signifying the greater thermal activation energy required for the larger scale of motion involved. The distinctive curvature in the Arrhenius plot of the $\alpha(T_g)$-process is exemplified in fig. 3.10(a) by the case of polyepichlorhydrin. At the higher temperature end the graph is roughly linear corresponding to an activation energy (ΔH^*) of 190 kJ mol^{-1} (0 to 6 °C), whereas at lower temperatures, towards T_g, ΔH^* appears to increase rapidly (430 kJ mol^{-1} at -20 °C). Fig. 3.10(b) shows that a good straight line is obtained in a WLF plot, though, confirming that the α-relaxation process near T_g is largely dependent on free volume. In some polymers the α-relaxation process catches up with the β-process at high temperatures on account of its steeper dependence on temperature, and the α- and β-peaks then tend to merge. However, they may be separated again on a temperature scan by applying high pressures, which greatly suppress the α-process, reflecting its requirement for free volume.

Molecular structure greatly affects glass transition temperatures and their associated dielectric relaxation times. Thus a bulky side group can

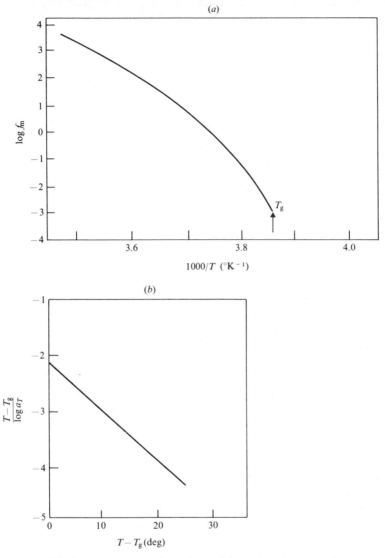

Fig. 3.10. Temperature dependence of the α-relaxation process in polyepichlor-hydrin: (*a*) Arrhenius plot, (*b*) WLF plot. Reprinted from Blythe and Jeffs (1969), p. 141, by courtesy of Marcel Dekker, Inc.

decrease T_g by preventing the chains from packing together tightly and vice versa. T_g can also be artificially reduced by adding a plasticiser; fig. 3.11 shows the effect on the α-relaxation of adding diphenyl to poly(vinyl chloride) (Fuoss, 1941).

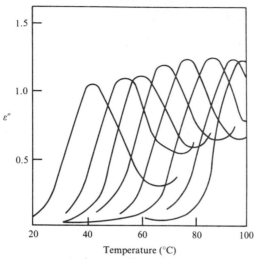

Fig. 3.11. The plasticising effect of diphenyl in poly(vinyl chloride): dielectric loss curves at 60 Hz for (right to left) 0, 1, 3, 6, 9, 12, 15, 20% diphenyl. Reprinted from Fuoss (1941), *J. Amer. Chem. Soc.*, **63**, 378. Copyright by the American Chemical Society.

When a polymer possesses no very polar groups, dielectric relaxations may be very weak effects which are scarcely observable, although the underlying molecular rearrangement processes will still be there. In these circumstances it is often possible to enhance the associated dielectric effect artificially by adding just a few polar groups which in other ways do not disturb the system too much. This technique has been most notably applied in the case of polyethylene, which is of great technological interest. Being non-polar its dielectric relaxations are especially weak, but by slightly oxidising the material, e.g. by milling in the presence of air, some of the $>CH_2$ groups are converted to polar $>C{=}O$ groups, which are readily accommodated even in the polyethylene crystal lattice. The molecular relaxation mechanisms can then be examined using dielectric methods, a valuable achievement, because the accessible frequency range of dielectric measurement is wider than that of any other kind. Various molecular processes in non-polar polymers may also show up accidentally as a consequence of a coupling of the motion of the polymer chains with the displacement of some small impurity molecules which may be polar or even ionic. By the same token a low concentration of a polar impurity may ruin the low dielectric loss of a non-polar polymer in a particular frequency region.

In partially crystalline polymers where crystalline and amorphous phases coexist in the solid, the relaxation spectrum becomes more complicated. Apart from the orientational processes taking part entirely in the amorphous regions, there are different mechanisms operating inside the crystals and at their boundaries. It is usually possible to decide whether a given loss peak belongs to the amorphous phase or is connected with crystals by varying the crystallinity. Thus reducing the crystallinity, by rapid quenching from the melt, enhances the strength of any relaxation process originating in the amorphous phase.

In crystals of polymers like polyethylene where the molecules are in their extended zigzag conformation, the molecular chains lie in a staggered arrangement with respect to their neighbours. If a chain has a polar group attached to it one can envisage that the application of an electric field will in general exert a couple tending to turn the chain about its axis, the magnitude of the couple depending on the angle between the dipole and the field. The repulsive forces from the surrounding chains in the crystal lattice will constrain chain rotation to jumps between two equilibrium positions, involving a 180° turn combined with a translational shift, as shown in fig. 3.12. On this model the molecules turn as rigid rods, so that we may expect the activation energy for the orientation process to be directly proportional to the length of chain which turns in the crystal. As a test of the above idea, dipolar relaxations of

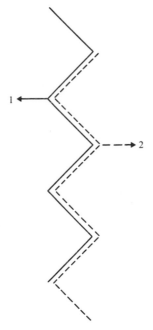

Fig. 3.12. Diagram of chain rotation in a polyethylene crystal.

ketones in paraffin waxes have been examined in cases where the ketone chain length is shorter than the thickness of the paraffin crystals, so that constraints on the ends of the rotating chains may be neglected. Experiment confirms the dependence of activation energy on the length of the dipolar chain (Meakins, 1962).

In the case of long polymer chains some direct evidence of the rotation of chains in crystals has been obtained from observations of the dependence of the relaxation strength on orientation of the chain axes with respect to the applied field. It is clear that rotation of chains which lie parallel to the field direction cannot contribute to the relaxation, and only a weak relaxation peak is found for crystals oriented in this way (Davies and Ward, 1969). If the chain-folds at the crystal boundaries were loose, a chain in the crystal would suffer little constraint on its rotation from outside the crystal, so that one might expect the activation energy for the relaxation to be directly proportional to crystal thickness (in the c-axis direction). For long chains, however, we must take into account the possibility of chain twisting, i.e. departure from rigid-rod behaviour. More elaborate theories suggest that twisting will develop as chains exceed 60–120 carbon atoms in length, when the activation energy should flatten out, causing the temperature of maximum dielectric loss (at a fixed frequency) to *saturate* with respect to chain length. Experimental data, though scanty, seem to suggest this view, as shown in fig. 3.13 (Hoffman, Williams and Passaglia, 1966).

Other complications may arise in dielectric relaxation spectra of

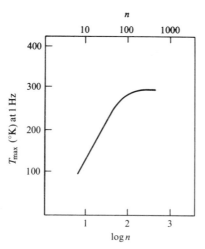

Fig. 3.13. The temperature of maximum dielectric loss at 1 Hz for crystalline long carbon-chain compounds T_{max} as a function of number n of CH_2 units in the chain. After Hoffman, Williams and Passaglia (1966).

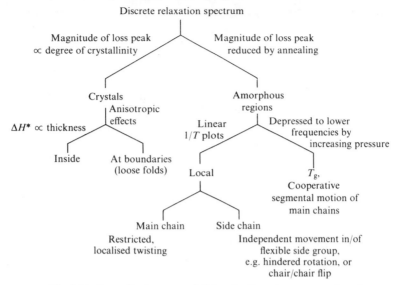

Fig. 3.14. Generalised pattern of dielectric relaxation processes in polymers.

polymers from chain branching, which may introduce a distinct relaxation process connected with molecular motion at a branch point, and from crosslinking which greatly restricts certain kinds of molecular movement.

Relaxation processes in solid polymers are summarised diagrammatically in fig. 3.14.

3.5   Liquids and solutions

End-over-end rotation of whole polymer molecules becomes a much more accessible mechanism for orientation of molecular dipoles when the material is in the liquid state, especially when it is dissolved in a low-molecular-weight solvent. With this additional feature in mind, there are three aspects of polymeric molecules which must be considered in order to reach an understanding of their dielectric behaviour in the liquid state: the type of dipole present on each repeat unit of the polymer, the equilibrium conformation of the individual molecules, and the flexibility of the molecular chains.

Polar groups in a polymer molecule may be classified according to the relative geometry of their dipole moment with respect to the contour of the polymer backbone. There are three main types:

(*a*) Dipole rigidly attached to the backbone and perpendicular to it.
(*b*) Dipole rigidly attached to the backbone and parallel to it.

(*c*) Dipole attached in a side chain which can move independently of the main chain.

Combinations of these types may also occur.

Perpendicular dipole units of type (*a*), which commonly occur in vinyl polymers, e.g. poly(vinyl chloride), can orientate individually by segmental motion, provided that rotations about the C–C bonds of the polymer chain are sufficiently free. We know that only in cases of exceptional steric hindrance or intramolecular hydrogen bonding do intramolecular barriers to such rotations lie above 20 kJ mol^{-1}, so that in liquids and solutions, where there is ample free volume available, we

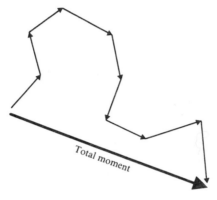

Fig. 3.15. Vectorial addition of unit dipoles along a polymer chain.

should expect dipolar relaxation to be very rapid and to be insensitive to the total length of the chain. This is borne out by experiment and a single relaxation process, which has a low activation energy and a relaxation time independent of the degree of polymerisation, is observed (North, 1972).

Parallel dipole units of type (*b*) add vectorially along the length of the polymer chain to give a cumulative moment which is proportional to the end-to-end distance of the molecule (fig. 3.15). This moment can only orient by end-over-end rotation of the whole molecule and remains unaffected by rotation of individual segments about the main chain axis. On this basis we predict a strong dependence of the associated relaxation process on molecular weight (the bigger the molecule the more difficult will it be for it to turn round), although we expect that the exact form of the dependence will be related to shape as well as to length of polymer chains (see below). This behaviour has been observed (Baur and Stockmayer, 1965) in liquid poly(propylene oxide). In this polyether the –CH$_3$

side groups disturb the normal geometry of the oxygen dipole and the dipoles on each monomer unit no longer bisect the C–O–C bond angle, so that each such dipole has a small component directed parallel to the contour of the main chain. The relaxation of the cumulative moment of these parallel components is seen as a subsidiary peak on the low-frequency side of the main loss peak, which is associated with relaxation of the perpendicular components by segmental motion (fig. 3.16). Increasing the molecular weight of the polymer shifts the subsidiary peak to a lower frequency.

A dipole moment of type (*c*) in a flexible side chain, e.g. as in poly(methyl methacrylate), imposes a dipole moment which can be

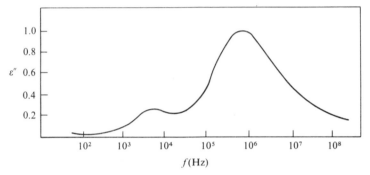

Fig. 3.16. Relaxation spectrum of liquid poly(propylene oxide) at $-30\,°C$ (Baur and Stockmayer, 1965).

resolved into a rigid, perpendicular component and a rotatable component. One might have supposed that two separate relaxation processes might then have occurred, but in practice only one, high-frequency, relaxation is observed and one concludes that in solution the side-group rotation and segmental motion in the main chain are combined in a single, fast process (North, 1972).

Because of the profound influence that the overall shape of a polymer chain undoubtedly has on any orientation process which involves rotation of the whole molecule, it is worthwhile digressing on the subject of equilibrium conformation. Flexible, linear polymers tend to be somewhat coiled-up in the liquid phase or in a solution. In a *good* solvent, polymer–solvent contacts are preferred energetically to polymer–polymer contacts and so the coils expand and the root-mean-square end-to-end distance, $(\overline{r^2})^{\frac{1}{2}}$, increases. On the other hand, in a *bad* solvent the polymer will contract and $(\overline{r^2})^{\frac{1}{2}}$ will decrease – ultimately phase separation will occur if the solvent is bad enough. At a certain temperature in a relatively poor solvent the effect of the intermolecular interactions will just offset the increase in end-to-end distance of the polymer due to excluded volume effects (occupation of space by a chain of finite thickness) and the polymer will then behave just like an unperturbed molecular chain of negligible thickness. The solvent in this ideal situation is called a *theta-solvent* and the chains adopt their *random-flight* or Gaussian conformation. It may be shown that the root-mean-square end-to-end distance $(\overline{r^2})^{\frac{1}{2}}$ is then proportional to the number of segments in the chain, and therefore to the molecular weight.

The frictional motion of deformable, random-coil molecules in solution has been examined theoretically in a general way. There are two basic models: the *free-draining* model (Rouse, 1953) which neglects hydrodynamic interaction between neighbouring parts of the same chain, and the *non-free-draining* model (Zimm, 1956) in which viscous drag between different segments assumes a dominant role. The latter model may be expected to give a better account of what happens inside the coils of a long molecule, and indeed, it gives good agreement with experimental viscosity data for dilute solutions of high polymers. The simpler Rouse model gives better agreement at higher concentrations, however, where the polymer molecules might be expected to intermingle. Presumably the hydrodynamic interaction between neighbouring segments of one molecule is nullified by the invasion of foreign segments.

The relaxation time τ for orientation of the total dipole moment of a chain, substantially by end-over-end rotation, but accompanied by some distortion, may be related to the viscosity of the solvent η and solution η_s on the basis of the above models:

$$\tau \text{ (free-draining)} = 1.21\frac{(\eta - \eta_s)}{cRT}M_w, \qquad (3.51)$$

$$\tau \text{ (non-free-draining)} = 0.85\frac{(\eta - \eta_s)}{cRT}M_w, \qquad (3.52)$$

where c is the concentration of the polymer solute and M_w is the molecular weight of the polymer. Although these equations do not give a great deal of physical insight into the orientation process they do predict fairly accurate values and give the correct molecular-weight dependence:

The backbones of some polymer molecules are of intermediate stiffness; they adopt a coil-like configuration in a solution, but their segmental mobility is so low that dipolar orientation is faster by whole molecule rotation than by twisting of individual dipole units. This behaviour has been observed in sulphones, e.g. poly(hex-1-ene sulphone) and poly(2-methyl pent-1-ene sulphone) in benzene and toluene (Bates, Ivin and Williams, 1967). Since the dipoles are rigidly attached to the main chain in these instances only one relaxation process is seen, and the relaxation time is molecular-weight dependent. Cellulose esters and ethers also fall into the stiff, coil-like class, but they generally show an additional high-frequency relaxation process, because they contain dipoles in rotatable side chains as well as dipoles fixed in the main chain. The high-frequency process in cellulose acetate is independent of molecular weight having a relaxation time of about 10^{-7} s at room temperature (Kuhn and Moser, 1963). This is appreciably longer than that in vinyl acetate, implying that in the cellulose acetate derivatives the acetate side groups are subject to considerable steric hindrance and do not benefit from cooperative backbone movement. The loss peak is no wider than that for a single Debye-like process, suggesting that each acetate group undergoes orientation independently.

Certain polymer molecules have a tendency to form rigid rods in solution. The best known examples are of biochemical origin, e.g. proteins, where the rigidity is brought about by intramolecular hydrogen bonding, although a few examples of polymers which adopt the rod form simply through excessive steric hindrance are also known. Theory indicates that the end-over-end rotational relaxation time of a rigid rod should vary as the cube of the rod length, so that the dielectric relaxation time for the axial component of the dipole moment should vary as the cube of the molecular weight. Such a dependence has been observed for dilute solutions of poly(n-butyl isocyanate) with molecular weights below 10^5 (Bur and Roberts, 1969). As a polymer chain becomes longer and longer, it becomes more and more difficult to sustain the rigid-rod configuration and some curvature is inevitable for very long chains. This will eventually lead, for sufficiently long chains, to random-coil configurations. The rotational relaxation time will then depend on the cube of the radius of gyration which varies as the square root of the chain length. We therefore expect a transition to a dependence of the dielectric relaxation time on the three-halves power of the molecular weight. This occurs in the case of poly(n-butyl isocyanate) when the molecular weight is above 10^5. At high molecular weight we might expect the rate of end-over-end rotational orientation to be overtaken by segmental orientation, implying a change from a stiff coil to a flexible coil. This has been observed in the case of poly (N-vinyl carbazole) (North and Phillips, 1968). In principle, we expect all polymers to be rod-like at very low molecular weights and to progress via stiff coils to flexible coils as the molecular weight is increased; this should be reflected in the dependence of dielectric relaxation time on molecular weight as indicated schematically in fig. 3.17.

Many biochemical polymer systems show polarisation effects attributable to migration of ion atmospheres surrounding the polymer chains. The mechanisms are then more akin to the behaviour of electrolytes than to dielectric materials and therefore will not be discussed further here.

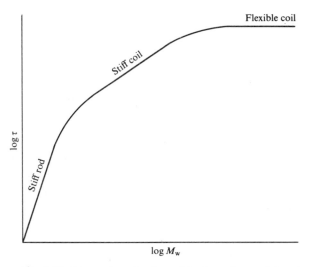

Fig. 3.17. Schematic molecular-weight dependence of the relaxation time for whole-molecule rotation.

3.6   Interfacial polarisation

So far in this chapter we have concerned ourselves with ideal specimens which are entirely homogeneous and connect perfectly with the electrodes which apply the electric field. In practice a material is always likely to have regions of non-uniformity, and impurities may be present as a second phase. Effects on dielectric properties attributable to material discontinuities are usually called Maxwell–Wagner effects, since Maxwell and Wagner were the first to consider them theoretically. Complications also often arise at electrodes where contact with the specimen may be incomplete and where entities like discharged ions may form spurious boundary layers. We shall see in this section how important it is to be aware of such anomalous effects; they can give totally misleading results if they are not recognised or avoided.

3.6.1   **Maxwell–Wagner effects**

A common form of discontinuity occurring in a solid dielectric is one
where cracks or voids are present. The dielectric constant will then
simply be reduced, depending on the amount and distribution of the
enclosed space or air. A much more serious type of heterogeneity,
however, is one where a relatively conductive component is mixed in an
insulator. Common instances of this occur where a material is contami-
nated with metallic particles or droplets of water. We can expect such
materials to behave like 3-dimensional RC-networks. On this basis we
would then predict an extra frequency-dependent contribution to dielectric constant and loss on account of the limited flow of current that is
possible within the isolated conductive regions.

Although Maxwell (1892) considered a rather special case when he
first examined the effect of discontinuities in dielectrics theoretically, he
obtained a very important result. He examined the effect of a field
applied across a specimen consisting of layers of two different materials
with dielectric constants ε_1', ε_2' and conductivities σ_1, σ_2 respectively. His
results showed that charges will accumulate in time at the interfaces
between the layers, whenever $\varepsilon_1'\sigma_2 \neq \varepsilon_2'\sigma_1$.

Wagner (1914) gave an approximate treatment of the important
practical case where a very highly insulating dielectric suffers from
inclusions of conductive impurities. Taking the model where the impur-
ity (dielectric constant ε_2', conductivity σ_2) exists as a sparse distribution
of small spheres (volume fraction f) in the dielectric matrix (dielectric
constant ε_1', negligible conductivity) he derived equations for the components of the complex dielectric constant of the composite:

$$\varepsilon' = \varepsilon_\infty'\left(1 + \frac{k}{1+\omega^2\tau^2}\right), \tag{3.53}$$

$$\varepsilon'' = \frac{\varepsilon_\infty' k\omega\tau}{1+\omega^2\tau^2}, \tag{3.54}$$

where

$$\varepsilon_\infty' = \varepsilon_1'\left[1 + \frac{3f(\varepsilon_2' - \varepsilon_1')}{2\varepsilon_1' + \varepsilon_2'}\right], \tag{3.55}$$

$$k = \frac{9f\varepsilon_1'}{2\varepsilon_1' + \varepsilon_2'}, \tag{3.56}$$

and

$$\tau = \frac{\varepsilon_0(2\varepsilon_1' + \varepsilon_2')}{\sigma_2}. \tag{3.57}$$

Comparing equations (3.53) and (3.54) with the Debye equations (3.25) and (3.26) we can appreciate that the composite displays a dielectric relaxation that is indistinguishable in form from that due to orientation of dipoles. Equation (3.57) shows that the relaxation time decreases as the conductivity of the material of the spheres increases, and the peak in tan δ may shift right up to radio-frequencies. For example, taking $\varepsilon_1' = \varepsilon_2' = 4$, and $\sigma_2 = 10^{-4}\,\Omega^{-1}\,m^{-1}$, we calculate a relaxation time of about 1 μs. Consequently, the effect can be very easily mistaken for one of dipole orientation, and great care must always be exercised in the interpretation of dielectric data when the presence of heterogeneities is suspected.

Sillars (1937) developed the subject further and he demonstrated the importance of the shape of conductive inclusions. The dielectric loss peak is enlarged and shifted to lower frequencies by any elongation in the direction of the applied field.

Experiments by Sillars (1937) and by Hamon (1953) using deliberately engineered heterogeneous systems have verified the theoretical predictions.

3.6.2   Electrode polarisation

Polarisation effects at electrodes become most prominent when the material of a specimen shows some appreciable bulk conductivity. Characteristically there is an apparent increase in the dielectric constant at low frequencies. The anomaly originates in a high-impedance layer on the electrode surface. This may be caused by imperfect contact between the metal electrode and the specimen, aggravated by the accumulation of the products of electrolysis etc. At low frequencies there is sufficient time for any slight conduction through the specimen to transfer all the applied field across the very thin electrode layers, and the result is an enormous increase in the measured capacitance. For a purely capacitive impedance C_e at the electrodes, in series with the specimen proper (geometrical capacitance C_0), Johnson and Cole (1951) showed that the apparent dielectric constant ε_{app}' takes the approximate form:

$$\varepsilon_{app}' = \varepsilon' + \frac{\sigma^2 C_0}{\omega^2 \varepsilon_0^2 C_e}, \tag{3.58}$$

where ε' and σ are the true (frequency-independent) dielectric constant and conductivity of the material of the specimen. This formula, which fairly accurately describes the behaviour observed with many liquids, can then be used to correct low-frequency measurements for electrode effects. For solids the electrode impedance is usually more complex, and since little detail is ever known about it, the problem of extracting meaningful results from low-frequency dielectric data is often rendered intractable by conduction.

3.7   Further reading

The original monographs by Debye (1929) and Fröhlich (1949) present the basic theory of dielectric relaxation. The books by Smyth (1955) and Hill *et al.* (1969) both give clear critical reviews of theoretical and experimental aspects. Results for solid polymers are most fully collected and discussed by McCrum, Read and Williams (1967) and by Hedvig (1977). North (1972) has reviewed dielectric relaxation in polymer solutions.

4 Measurement of dielectric properties

4.1 Introduction

One inherent advantage of studying materials through their electrical properties is that electrical measurements can be made over a very wide range of frequencies with a high degree of precision. Thus the dielectric constant and loss of solid polymers are often known from 10^{-4} Hz up to optical frequencies. Taken together with the possibility of varying temperature and pressure, the breadth of the experimental technique becomes very impressive, and researchers have made full use of this tool in probing underlying molecular behaviour. The detailed way in which dielectric properties may be mapped is well exemplified by the data for poly(ethylene terephthalate). Fig. 4.1 shows relief models of the variation of dielectric constant and loss with temperature and frequency for this polymer (Reddish, 1962).

In considering the methods of measurement we must distinguish two basic types, namely *lumped-circuit* and *distributed-circuit* methods. The aim of lumped-circuit methods, which are always used at lower frequencies, is to determine the equivalent electrical circuit of a specimen at a given frequency. From the previous discussion in § 3.1 of the current-voltage relationship for a dielectric material between the plates of a capacitor, we can see directly how the components of the complex permittivity of the material may be expressed in terms of an equivalent parallel circuit. Thus let the specimen be represented by a capacitance C_p in parallel with a resistance R_p, as shown in fig. 4.2(a). The total impedance Z will then be given by

$$\frac{1}{Z} = \frac{1}{R_p} + i\omega C_p. \tag{4.1}$$

Application of the alternating voltage represented by the real part of $V = V_0 e^{i\omega t}$ will produce an out-of-phase or capacitive current I_C:

$$I_C = \text{Imaginary part} \left[\frac{V}{Z} \right]$$
$$= i\omega C_p V, \tag{4.2}$$

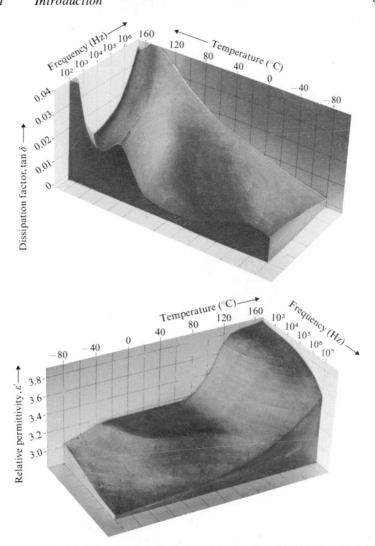

Fig. 4.1. Photographs of relief models showing the variation of dielectric constant and loss of poly(ethylene terephthalate) with temperature and frequency of measurement (Reddish, 1950).

and an in-phase or resistive current I_R:

$$I_R = \text{Real part} \left[\frac{V}{Z} \right]$$

$$= \frac{V}{R_p}. \tag{4.3}$$

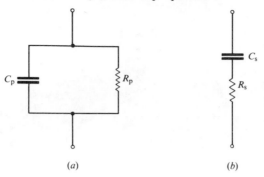

Fig. 4.2. Equivalent electrical circuits of dielectric specimens: (*a*) parallel, (*b*) series.

Comparing equations (4.2) and (4.3) with equations (3.8) and (3.9), respectively,

$$\varepsilon' = \frac{C_p}{C_0},$$ (4.4)

$$\varepsilon'' = \frac{1}{R_p C_0 \omega},$$ (4.5)

and

$$\tan \delta = \frac{\varepsilon''}{\varepsilon'} = \frac{1}{R_p C_p \omega}.$$ (4.6)

Alternatively, the specimen may be regarded in terms of a series circuit. If the equivalent series components of capacitance and resistance are C_s and R_s, respectively (see fig. 4.2(*b*)), the total impedance will be given by

$$Z = R_s + \frac{1}{i\omega C_s}.$$ (4.7)

If we again compare out-of-phase and in-phase currents after application of the alternating voltage, we obtain

$$\varepsilon' = \frac{C_s}{C_0(1 + \tan^2 \delta)},$$ (4.8)

$$\varepsilon'' = \frac{R_s C_s}{C_0(1 + \tan^2 \delta)},$$ (4.9)

and

$$\tan \delta = \frac{\varepsilon''}{\varepsilon'} = R_s C_s \omega.$$ (4.10)

We can therefore calculate the dielectric constant and loss of a material from measured values of either equivalent series or parallel circuit components of a specimen.

The reciprocal of the specimen resistance in the equivalent parallel circuit for a given frequency is sometimes called the AC conductance G_p. It is a combination of DC conductance, by which we mean any real flow of charge through the sample under the influence of the applied field, and the *anomalous* conductance due to any time-dependent polarisation processes. The contribution that a true DC conductivity σ (conductivity is defined as the reciprocal of resistivity which is the resistance between opposite faces of a unit cube of material) will make to the dielectric loss at an angular frequency ω can be readily calculated as follows for the material in a parallel-plate capacitor. Substituting for resistance in equation (4.5):

$$\varepsilon'' = \frac{G_p}{C_0\,\omega}. \tag{4.11}$$

If the capacitor plates have area A and separation s:

$$G_p = \frac{\sigma A}{s} \quad \text{and} \quad C_0 = \frac{\varepsilon_0\,A}{s}, \tag{4.12}$$

and hence

$$\varepsilon'' = \frac{\sigma}{\varepsilon_0\,\omega}. \tag{4.13}$$

This shows how DC conductivity causes ε'' to rise rapidly at low frequencies.

At high frequencies the electromagnetic wavelength inevitably becomes comparable with sample dimensions, and lumped-circuit methods must be abandoned in favour of distributed-circuit methods in which the sample becomes the medium for propagation of electromagnetic waves. Dielectric constant and loss must then be obtained from the observed wavelength and attenuation characteristics.

As an alternative to methods which use an applied sinusoidal voltage, called *frequency-domain* measurements, a step change in voltage across a specimen may be made and the ensuing current transient observed. Such time-domain measurements may be translated into frequency-domain terms by Fourier integral transformations.

The methods which are appropriate for the various frequency regimes are charted in fig. 4.3.

There are some general principles concerning the design of specimen holders for lumped-circuit methods which should be discussed before

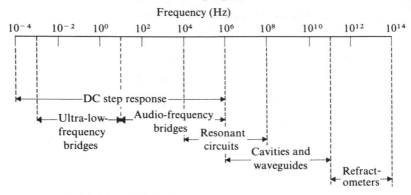

Fig. 4.3. Chart of dielectric measurement methods.

going on to describe individual measurement methods. Most specimen holders are based on a simple parallel-plate capacitor, taking a thin disc of solid material between their electrodes or being filled with liquid.

The primary drawback of a parallel-plate capacitor is the fringing of the field at the edges (fig. 4.4(*a*)). Whatever the arrangement of a 2-electrode system, it is difficult to make an accurate edge correction, because it changes in the presence of the test material. Semi-empirical corrections have been derived for various arrangements but it is better, if possible, to avoid the difficulty altogether by using a guard electrode as shown in fig. 4.4(*b*). The guard is held at the same potential as the guarded electrode but is not connected to it. Provided the specimen and guard electrode extend beyond the guarded electrode by at least twice the specimen thickness, and provided the guard gap is small in comparison with the specimen thickness, the field distribution in the guarded area is identical with the specimen in place and with nothing between the electrodes; the dielectric constant is the direct ratio of the capacitances

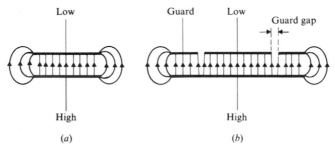

Fig. 4.4. Electric field patterns in parallel-plate electrode systems: (*a*) un-guarded, (*b*) guarded.

between the guarded electrode and the counter electrode in the two cases. Furthermore, the geometry is defined by the guard electrode so that the vacuum capacitance may be accurately calculated from the electrode dimensions. The guard electrode may also serve to eliminate the effect of surface conduction round the edges of the specimen, which can be a nuisance with hygroscopic materials.

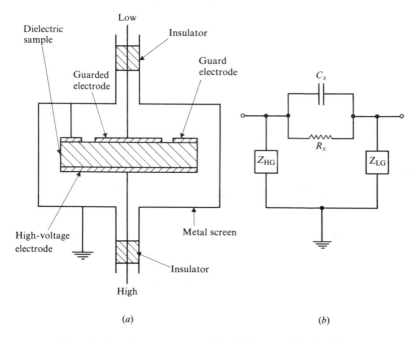

Fig. 4.5. A 3-terminal specimen holder: (*a*) diagram of shielded electrode system with guard ring, (*b*) equivalent electrical circuit of the cell.

Stray capacitances, e.g. between a lead and the opposite electrode, must be avoided by proper shielding. For this reason it is preferable to run the guarded electrode at zero potential. Then the guard electrode can be earthed and a completely shielded 3-terminal cell constructed (fig. 4.5), where the only stray capacitances are between the electrodes and earth, giving an equivalent circuit as shown.

Unfortunately, with most high-frequency measurement methods it is not possible to use a guarded electrode system and, in addition, lead inductances become appreciable at high frequencies. To meet the high-frequency requirement a special 2-electrode system, as shown in fig. 4.6, has been developed to eliminate errors due to lead impedances and to reduce edge effects as much as possible. (A built-in side capacitor is also

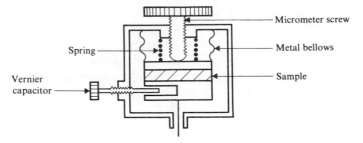

Fig. 4.6. Diagram of a 2-terminal micrometer electrode cell.

provided for use in resonance methods – see § 4.3.) The cell capacitance is first measured with the specimen clamped between the electrodes. The specimen is then removed and the electrodes adjusted by means of the micrometer to give the same capacitance. The edge correction is reduced to a minimum by using a specimen of the same size as the electrodes. For accurate capacitance measurements the edge effect may be reduced even

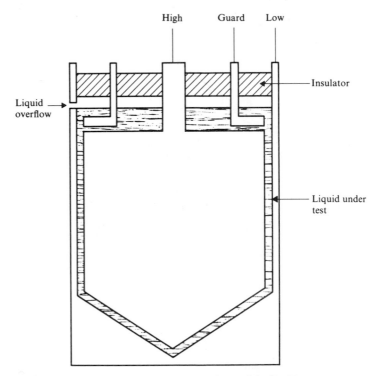

Fig. 4.7. Diagram of a 3-terminal electrode cell for liquids.

further by immersing the specimen in a fluid whose dielectric constant is approximately matched to that of the samples.

Any air gaps between the cell electrodes and specimen surfaces will introduce a series capacitance. Accidental gaps are usually avoided by applying thin foil electrodes, e.g. tin foil applied with a minimum quantity of pure Vaseline or silicone grease, or by evaporating metal on to the specimen. Alternatively, when very low losses are to be measured and it is desirable not to introduce any foreign matter into the sample, a known air gap may be deliberately left above the specimen and allowed for in the calculation of specimen capacitance and loss.

Liquid cells are designed along the same lines as those for solids. A typical 3-terminal version is shown in fig. 4.7.

4.2   Bridge methods

The most widely used way of determining the equivalent capacitance and resistance of a specimen is to use a type of Wheatstone bridge network to compare the unknown with standard components. The most versatile form for measurements of dielectric materials is the Schering bridge with which it is possible to make very accurate measurements over the frequency range 10 to 10^5 Hz (the audio-frequency range). The basic circuit for a practical form of the bridge is shown in fig. 4.8. At balance, when the detector indicates a null, we have the usual relation between the impedances of the arms:

$$Z_1/Z_2 = Z_3/Z_4. \tag{4.14}$$

Equal resistive ratio arms are used and a small variable capacitor is applied across resistance 1 to balance the conductance of the unknown in arm 4. A substitution procedure is generally used. First, the bridge is

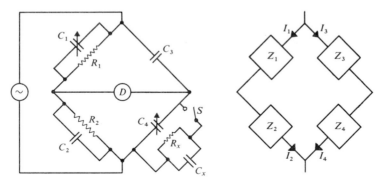

Fig. 4.8. Circuit diagrams of a conjugate Schering bridge.

balanced with the sample cell *out* (switch *S* open). Then the sample cell is connected *in* (switch *S* closed) and balance regained by compensating for the cell capacitance C_x by decreasing C_4, and for the cell conductance G_x by increasing C_1. From the two sets of balance conditions we obtain, for $\tan \delta_x < 10^{-2}$,

$$C_x = (C_4)_{out} - (C_4)_{in} = \Delta C_4, \qquad (4.15)$$

$$\tan \delta_x = \frac{G_x}{\omega C_x} = \frac{\omega R_1 (C_4)_{out} \Delta C_1}{C_x}. \qquad (4.16)$$

The success of an audio-frequency bridge depends on being able to deal with stray impedances which couple various parts of the bridge by unknown amounts, depending on component settings and the position of the operator. Thus the foregoing balance conditions, (4.15) and (4.16), really only hold true when $I_1 = I_2$ *and* $I_3 = I_4$, which means that there must be no stray capacitances from the detector terminals to earth. From this point of view the Schering bridge is particularly good, because almost complete shielding of the bridge components from one another may be achieved. Firstly, the absence of series connections in any bridge arm between capacitance and resistance components simplifies the shielding. Secondly, a voltage divider may be connected across the generator terminals to ensure that the detector terminals are at earth potential at balance, as shown in fig. 4.9. With this arrangement, called a Wagner earth, the modified measurement procedure is as follows. The bridge is first balanced as normal, and then the detector connected from one bridge corner to earth and the Wagner circuit adjusted to give another null. This disturbs the bridge slightly which is then rebalanced and so on until, simultaneously, the bridge is balanced and the detector terminals are at earth potential. In this condition, earthed screens can be used around all bridge components, and a guard electrode can be used in the sample cell.

Using these techniques great precision can be achieved with a Schering bridge over all of its working range. Differences in δ as small as 300 nrad may be resolved (Astin, 1936).

Another kind of audio-frequency bridge, which has come to the fore recently and which is very convenient to use, is the transformer ratio-arm-bridge, shown schematically in fig. 4.10. This bridge uses inductive ratio arms to compare the unknown directly with standard components. The voltage transformer is energised by an oscillator connected to its primary winding and develops one voltage V_1 between the unknown impedance Z_u and the neutral (earthed) line, and a second voltage V_2, 180° out of phase with the first, across the standard Z_s. The currents I_1

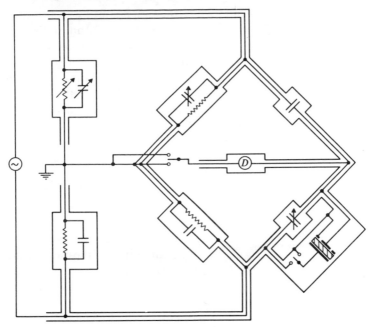

Fig. 4.9. Circuit diagram of a shielded conjugate Schering bridge with a Wagner earth and a guarded specimen holder.

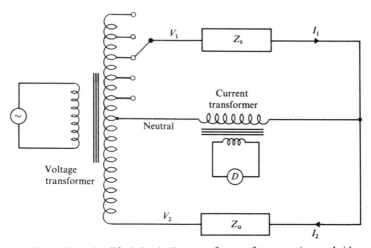

Fig. 4.10. A simplified circuit diagram of a transformer-ratio-arm bridge.

and I_2 flowing through Z_u and Z_s are then V_1/Z_u and V_2/Z_s, respectively. When these currents are equal they combine, on account of their phase difference, to produce zero core flux in the current transformer, and the detector indicates a null. The circuit has two big advantages: (*a*) impedances between the unknown and earth merely shunt the low resistances of the voltage and current transformers and do not affect the bridge balance, so that long screened leads and guarded electrodes can be used; (*b*) the voltage transformer can be tapped very accurately to obtain decade ratios, so that only a few standards are required. The principal drawback is that full sensitivity is restricted to the small frequency range for which the transformers are designed.

Below about 20 Hz transformers become very inefficient and more specialised bridges are necessary. Below about 10^{-2} Hz bridge balancing also becomes a very slow business, because the period of each cycle is so long that one has to wait a long time after each readjustment of the bridge to see the resulting change in the amplitude of the output signal at the detector. The frequency range can be extended down to about 10^{-3} Hz, however, by balancing the bridge during just part of a cycle using a phase-sensitive detector.

At high frequencies, bridges can still be used provided that special precautions are taken to eliminate the effects of stray inductances which become very large, but it is usually better to adopt a resonance method above 10^6 Hz.

4.3   Resonance methods

At radio frequencies a medium- to low-loss material can be most sensitively examined by making it part of a resonant circuit. The Hartshorn and Ward (1936) method, for which the basic circuit is shown in fig. 4.11, has been highly developed (Reddish *et al.*, 1971) for very accurate measurements in the 10^5 to 10^8 Hz range. Here a disc specimen is held in a micrometer electrode system (see fig. 4.6, where the specimen is represented by a capacitance C_x in parallel with a resistance R_x) and connected directly to a coil of fixed inductance L. The circuit is energised by a loosely coupled oscillator whose frequency is tuned to resonance as indicated by the voltmeter, and the maximum voltage V_i noted. The specimen is now removed, the circuit retuned to resonance by decreasing the inter-electrode gap by an amount Δx, and the new maximum voltage V_0 noted. Finally, with the aid of the side capacitor C, the capacitance width ΔC_0 of the resonance curve at half-power height is determined with the specimen absent. Since the circuit capacitance must be the same

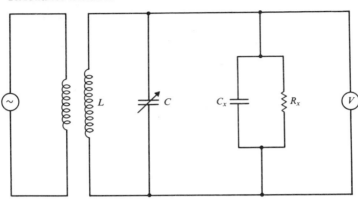

Fig. 4.11. The basic electrical circuit of the Hartshorn and Ward (1936) resonance method of dielectric measurement.

for the two resonance conditions, we obtain immediately for the dielectric constant of the material

$$\varepsilon' = \frac{t}{t - \Delta x}, \tag{4.17}$$

where t is the specimen thickness. This ignores any change in edge capacitance due to the presence of the specimen. Although some correction may be made for this, a more reliable alternative is to use a liquid immersion cell if the ultimate in accuracy is wanted.

Circuit analysis shows that the loss tangent of the material is given by

$$\tan \delta = \left(\frac{V_0}{V_i} - 1\right) \frac{\Delta C_0}{2C_0}, \tag{4.18}$$

where C_0 is the calibrated air capacitance of the main electrode set at resonance with the sample out. In its most refined form the method is capable of measuring a δ of 50 μrad to ± 1 μrad.

Measurements may be made at different frequencies by changing the inductance coil. In going to high frequencies the required inductance eventually becomes impractically small, however, and stray inductances take over. It is then necessary to use a resonant cavity of some kind, and a re-entrant cavity of the type shown in fig. 4.12 is often used in the 10^8 to 10^9 Hz range. The circuit is a hybrid one in the sense that its inductance and fixed capacitances are distributed along a short-circuited, coaxial transmission line, whereas the electrode gap in the centre conductor forms a lumped, variable capacitor, which may be calibrated in the normal way in air at low frequencies.

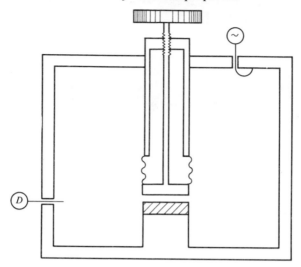

Fig. 4.12. Diagram of a re-entrant cavity for dielectric measurements.

Measurement with the re-entrant cavity is similar to the above Hartshorn and Ward method, except that a decrease in the electrode gap necessarily incurs an increase in inductance, and this must be taken into account (Works, 1947). The latter effect may be expressed as an equivalent capacitance change ΔC_L, and, assuming that the inductance increases like that of a loss-free transmission line, we may calculate that

$$\frac{dC_L}{dl} = \frac{1}{cZ_0} \operatorname{cosec}^2 \frac{2\pi l}{\lambda}, \tag{4.19}$$

where l is the length of the line, Z_0 its characteristic impedance, λ the wavelength, and c is the velocity of light. The capacitance width of the resonance curve with the sample absent may be determined by fine adjustment of the electrode gap, again taking into account the accompanying change in inductance. The $\tan \delta$ value is then calculated from equation (4.18).

4.4　Wave-transmission methods

As the frequency is taken above about 10^9 Hz, electromagnetic wavelengths become comparable with typical specimen dimensions. Applied fields then vary from place to place within a specimen and we must analyse the dielectric response in terms of Maxwell's wave equations. Experimentally, it is convenient to confine the waves by using a coaxial transmission line, consisting of a central conductor enclosed in a hollow conducting tube, or, at very high frequencies, a simple waveguide of rectangular or circular cross-section. The wavelength range over which this technique can be used, approximately 1 to 300 mm, is normally called the microwave region.

A preliminary step to dielectric measurement by wave-transmission techniques is to relate the basic wave parameter, called the propagation factor γ^*_s of the material, to permittivity. In terms of the propagation factor the equations for the electric and magnetic fields of a plane wave travelling in the x-direction in a uniform, infinite material are

$$E = E_0 \exp{(i\omega t - \gamma^*_s x)} \tag{4.20}$$

$$H = H_0 \exp{(i\omega t - \gamma^*_s x)}, \tag{4.21}$$

where the respective field vectors E and H are orthogonal (a transverse electromagnetic, or TEM, mode). In accordance with Maxwell's electromagnetic field equations the complex propagation factor of the material is given by:

$$\gamma^*_s = i\omega(\varepsilon^*_s \mu^*_s)^{\frac{1}{2}} = \alpha_s + i\beta_s, \tag{4.22}$$

where ε^*_s and μ^*_s are the absolute (complex) permittivity and permeability of the material. The real part α_s defines the attenuation (zero in a vacuum) of the wave, and the imaginary part β_s defines the wavelength λ_s in the material:

$$\lambda_s = \frac{2\pi}{\beta_s}. \tag{4.23}$$

In a vacuum the propagation factor simplifies to

$$\gamma_0 = i\omega(\varepsilon_0 \mu_0)^{\frac{1}{2}} = i\beta_0, \tag{4.24}$$

where ε_0, μ_0 are the permittivity and permeability of free space, respectively. The complex relative permittivity of a non-magnetic ($\mu^* = \mu_0$) dielectric material at angular frequency ω is then related to the propagation factor of a plane wave of that frequency travelling through the material by

$$\varepsilon^* = -\gamma^{*2}_s \left(\frac{\lambda_0}{2\pi}\right)^2, \tag{4.25}$$

where λ_0 is the free-space wavelength at that frequency. Equating real and imaginary parts:

$$\varepsilon' = \frac{\lambda_0^2}{\lambda_s^2}\left(1 - \frac{\alpha_s^2}{\beta_s^2}\right) = n^2\left(1 - \frac{\alpha_s^2}{\beta_s^2}\right), \tag{4.26}$$

$$\varepsilon'' = \frac{\lambda_0^2}{\lambda_s^2} \cdot \frac{2\alpha_s}{\beta_s} = n^2 \frac{2\alpha_s}{\beta_s}, \tag{4.27}$$

n being the refractive index of the material at the particular frequency concerned. Taking into account boundary conditions, it may be shown that equations (4.26) and (4.27) apply not only to plane waves propagating in the body of an isotropic material, but also to the principal (or TEM) mode of propagation in a coaxial transmission line. For all modes in other waveguides or for higher order modes along a coaxial line, however, the basic relation (4.25) must be modified as follows (von Hippel, 1954):

$$\varepsilon^* = \frac{(1/\lambda_c)^2 + (\gamma_{sg}^*/2\pi)^2}{(1/\lambda_c)^2 + (1/\lambda_{og})^2}, \tag{4.28}$$

where γ_{sg}^* is the propagation factor of the particular mode in the guide filled with the material in question, λ_{og} is the mode wavelength in vacuum, and λ_c is the cut-off wavelength of the guide. For a wave to pass along a waveguide its wavelength must be less than a certain critical value, the cut-off wavelength, which is related to the cross-sectional dimensions of the guide. If the dimensions are chosen correctly, transmission can be limited to a single transverse mode TE_{01}. In a coaxial line only the principal TEM mode (for which $\lambda_c = \infty$) will propagate, provided the wavelength is above a critical value, which is of the order of the separation between the inner and outer conductors. Clearly, for measurement purposes, multiple modes must be excluded and consequently practical frequency ranges for the application of coaxial-line and waveguide methods are 10^8 to 5×10^9 Hz and 3×10^9 to 6×10^{10} Hz, respectively.

In the basic method of Roberts and von Hippel (1946), standing waves are established in a waveguide or coaxial line with a closed (shorted) end and with the dielectric sample filling the space inside to a distance d from the closed end. A diagram of a suitable apparatus is shown in fig. 4.13. If the sample is a liquid it is necessary, of course, to mount the tube vertically. The standing wave takes the form shown in fig. 4.14. The wave in the space in front of the samples is characterised by measuring (*a*) the inverse standing wave ratio E_{min}/E_{max}, i.e. the ratio of the standing wave amplitudes at node and antinode positions, (*b*) the distance x_0 of the first node from the surface of the sample, and (*c*) the wavelength λ_{og}, given by twice the separation of adjacent nodes. This is done by means of a small travelling probe inserted into the guide

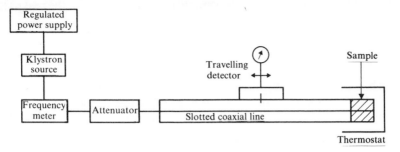

Fig. 4.13. Diagram of a slotted-line apparatus.

through a longitudinal slot. Roberts and von Hippel were able to show that the propagation factor γ^{*}_{sg} of the wave in the sample is related to the experimental observables in the following way:

$$\frac{\tanh \gamma^{*}_{sg}d}{\gamma^{*}_{sg}d} = -i\frac{\lambda_{og}}{2\pi d} \cdot \frac{E_{min}/E_{max} - i\tan(2\pi x_0/\lambda_{og})}{1 - i(E_{min}/E_{max})\tan(2\pi x_0/\lambda_{og})}. \qquad (4.29)$$

Numerical procedures have been devised for solving this complex transcendental equation.

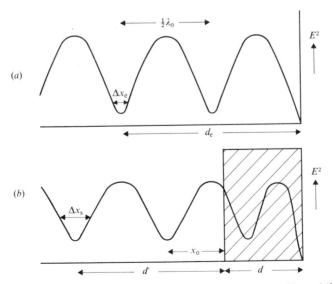

Fig. 4.14. Diagram of standing waves (a) in an empty waveguide, and (b) with a dielectric specimen in place at the closed end.

When the dielectric loss of the sample is low ($\tan \delta < 0.1$), the real part of equation (4.29) reduces to

$$\frac{\tan \beta_s}{\beta_s} = -\frac{\lambda_{og}}{2\pi d} \tan \frac{2\pi x_0}{\lambda_{og}}, \tag{4.30}$$

which may be solved by reference to tables of $(\tan \theta)/\theta$. For a coaxial line ($\lambda_{og} = \lambda_0$) we may then obtain the following expressions for the dielectric constant and loss of the sample:

$$\varepsilon' = \left(\frac{\beta_s \lambda_0}{2\pi}\right)^2, \tag{4.31}$$

$$\tan \delta = \frac{E_{min}}{E_{max}} \cdot \frac{\lambda_0}{\pi d} \cdot \frac{\beta_s \delta[1 + \tan^2(2\pi x_0/\lambda_0)]}{\beta_s d(1 + \tan^2 \beta_s d) - \tan \beta_s d}.$$

The experimental procedure may also be modified for a low-loss sample. It may be shown that the inverse standing wave ratio may be determined very accurately by measuring the width of a node:

$$\frac{E_{min}}{E_{max}} = \frac{\pi \Delta x}{\lambda_0}, \tag{4.33}$$

where Δx is the distance between double-power points on either side of a minimum. When the loss due to the sample *is* very low, it also becomes necessary to take into account wall losses in the transmission line. The width Δx_e of a node distance d_e from the end of the empty line is a measure of these, and may be used to correct the results. First the observed width Δx_s of a node distance d' in front of the sample is corrected for the wall losses in the empty section in front of the sample:

$$(\Delta x_s)_{\text{corrected}} = (\Delta x_s)_{\text{observed}} - \frac{d'}{d_e}\Delta x_e. \tag{4.34}$$

The total $\tan \delta$ value for the sample-filled section is then calculated by equation (4.32) and the wall-loss component $\Delta x_e/d_e$ (assuming $\mu = \mu_0$) subtracted to give the net value for the sample alone.

4.5 Time-domain methods

Sudden application of a steady voltage across a capacitor containing a dielectric will produce a transient charging current whose form depends on the rate at which equilibrium polarisation is attained in the dielectric. In principle, the equivalent point-by-point *frequency-domain* behaviour of the dielectric can then be obtained from this time-dependent response by a Fourier integral transformation which extracts all the separate harmonic components from the one transient signal in the *time domain*. For a linear dielectric in a capacitor (vacuum capacitance C_0) subjected to a step voltage V_0 at time $t=0$, the complex dielectric constant is given in general by

$$\varepsilon^*(\omega) = \frac{1}{C_0 V_0} \int_0^\infty I(t) e^{-i\omega t} \, dt, \qquad (4.35)$$

where $I(t)$ is the transient charging current at time t. In practice, the current transient will consist of an initial pulse corresponding to the atomic and electronic polarisations, which are effectively instantaneous, followed by a decaying current from any relaxing, dipolar orientation processes. If the dielectric is appreciably conductive (DC conductance G), there will also be a constant current component. We may therefore rewrite equation (4.35) as

$$\varepsilon^*(\omega) = \varepsilon_\infty + \frac{1}{C_0 V_0} \int_0^\infty I(t) e^{-i\omega t} \, dt - i \frac{G}{\omega C_0}, \qquad (4.36)$$

where the integration now excludes the initial current pulse and the constant conduction current. The form of current and charge transients for simple step voltage changes are shown in fig. 4.15, where

$$q = \int_0^t I \, dt.$$

The complete Fourier transformation of current-transient data involves very extensive numerical computation and two approximate methods have often been used instead. The first method, known as the Hamon (1952) approximation, is based on the assumption that the current transient conforms to the equation

$$I(t) = A t^{-n} \quad (0.5 < n < 1), \qquad (4.37)$$

which is indeed found to be true experimentally over limited portions of a transient curve. Hamon was able to show that the dielectric loss was in that case given approximately by the simple formula

$$\varepsilon^*(\omega) = \frac{I(t)t}{0.63 \, C_0 V_0}, \qquad (4.38)$$

with $\omega t = \pi/5$. This result, which reflects the fact that the main contribution to the Fourier transform for frequency ω is related to behaviour of the current transient at times near $t = 1/\omega$, is quite accurate for typical Debye-like relaxation processes in polymers.

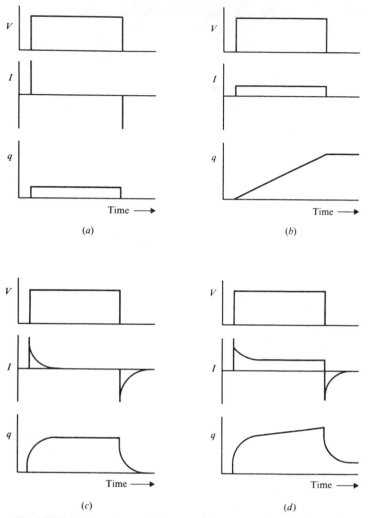

Fig. 4.15. Transient currents from step voltages across (*a*) a loss-free dielectric, (*b*) a conductor, (*c*) a dielectric with dipolar relaxation, and (*d*) a dielectric with dipolar relaxation and conduction.

A better alternative has recently been given by Hyde (1970) who bases his method on the experimental fact that the rise in the time integral $q(t)$ of the transient current is always a gradual one with respect to log t. Accordingly, an economic record of the transient over a wide frequency range can be obtained by measuring q at equal intervals on a logarithmic time scale. Considering the intrinsic width of a Debye-like relaxation process (frequency half-width$=1.14$ decades) a log 2 interval (i.e. $t_{n+1}/t_n=2$) between data points gives adequate time resolution, and is convenient for data processing on a computer. Defining a general time interval, $2^{n-\frac{1}{2}}t_1$, to $2^{n+\frac{1}{2}}t_1$, over which the increment in charge is $\Delta q(n)$, and choosing $t_1=1/\omega$, the fast Fourier transforms then take the forms:

$$\varepsilon'(\omega)=\sum_{p=-\infty}^{+\infty}\Delta q(n)x(n), \tag{4.39}$$

$$\varepsilon''(\omega)=\sum_{p=-\infty}^{+\infty}\Delta q(n)y(n), \tag{4.40}$$

where $n=2^p$ and $x(n)$ and $y(n)$ are constant coefficients for the given frequency ω. In practice only the first few terms of the transforms are needed: just five terms are sufficient for 1% accuracy in $\varepsilon^*(\omega)$.

In the past the DC step-response technique has been used primarily to measure dielectric constant and loss at very low frequencies (down to 10^{-4} Hz) which were not accessible in any other way. The basic experimental technique is very simple. A step voltage in the range 1 to 500 V is applied across a sample held in a normal guarded-electrode system and the resulting charging current is detected by the voltage which it produces across a standard resistor in the circuit. For the very small transient polarisation currents normally encountered, typically in the 10^{-12} to 19^{-9} A range, a very high-value standard resistor (10^9 to $10^{12}\,\Omega$, say) has to be used. Since the input capacitance C_i of the electrometer used to measure the voltage is usually about 10 pF and since it shunts the standard resistor, the time constant ($=R_sC_i$) for the response of the measuring circuit is long – about 10 s for a resistance of $10^{12}\,\Omega$. This means that transients cannot be directly measured at short times after application of the step voltage by the simple technique. Consequently, a circuit like that shown in fig. 4.16(a) is often used. Here the electrode connected to the electrometer amplifier is virtually kept at earth potential by negative feedback, which counteracts the voltage developed across the standard resistor R_s by the current i flowing through it. The input capacitance of the amplifier no longer shunts R_s and the time constant is reduced by a factor equal to the amplifier gain N, which may be 1000 or more. Then at times long in comparison with the effective time constant (i.e. $t\gg R_xC_i/N$) the polarisation current is given by

$$I=-V/R_s, \tag{4.41}$$

where V is the output voltage of the amplifier. The discharging current is also recorded as the voltage is reduced to zero by switching the upper electrode to earth.

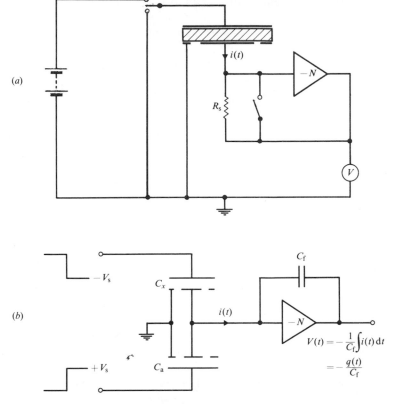

Fig. 4.16. Electrical circuits for DC step-response methods: (a) measurement of a current transient, (b) measurement of the time integral of current, with compensation for the large instantaneous component of the response.

In a more recent version of the method the time integral of the current is measured by an electrometer amplifier with negative capacitive feedback, and the required dynamic range of the measuring circuit is much reduced by backing off the instantaneous response of the sample with a signal from an equivalent air capacitor fed with a voltage step of opposite sign (see fig. 4.16(b)). This technique has been extended, using in-line computation, to provide a very rapid means of making measurements of dielectric constant and loss over the frequency range 10^{-4} to 10^6 Hz (Hyde, 1970).

The Fourier integral transform technique has also been used recently to obtain very high frequency (10^8–10^{10} Hz) data. The experimental method, usually referred to as time-domain spectroscopy, involves subjecting a sample in a waveguide to a train of very short radiation pulses, with fast rise times, and then observing the reflected or transmitted pulses with a sampling oscilloscope (Suggett, 1972).

4.6   Further reading

Measurements of dielectric properties by bridge, resonance and wave-guide methods are described in the basic text edited by von Hippel (1954). Much practical information, including detail on the design of sample holders and the preparation of specimens, is also given in national publications of standard test methods. The most relevant document from the British Standards Institution is *Methods of Testing Plastics. Part 2, Electrical Properties* (BS2782:1970), and that from the American Society for Testing and Materials is the section on 'Electrical Insulating Materials' in the current issue of the *Annual Book of ASTM Standards*.

5   Conduction in polymers

5.1   Introduction

Electrical conductivity of materials is a property which spans a very wide range as may be judged from the conductivity chart (fig. 5.1). Organic compounds typically have conductivities eighteen orders of magnitude smaller than those of metals, and the polymeric subgroup falls at the low-conductivity end, with polyethylene, polytetrafluoroethylene and polystyrene being amongst the best insulators known. In contrast, metallic superconductors have immeasurably high conductivities in their low-temperature, superconducting regimes.

Electrical conduction may occur through the movement of either electrons or ions. In each case, however, a suitable starting point for discussion of the conduction process is the basic equation

$$\sigma = qn\mu, \tag{5.1}$$

where the conductivity σ is resolved into three factors: the charge q, concentration n and drift mobility μ of the carriers. The latter parameter characterises the ease with which the charged species will move under the influence of the applied electric field and is usually expressed as a velocity per unit field ($m^2 V^{-1} s^{-1}$). There may be contributions to the conductivity from several different types of carrier, notably electrons and holes (a hole is an electron vacancy carrying an equivalent positive charge) in electronic conductors, and cation and anion pairs in ionic conductors. Theories of conduction aim to explain how n and μ are determined by molecular structure and how they depend on such factors as temperature and applied field.

In most polymeric materials it is very difficult to observe any electronic conductivity at all, and, what conductivity there is, usually depends on the movement of adventitious ions. Consequently, any improvements in the quality of insulation are generally won by careful preparation and purification, so as to avoid as much as possible the presence of ionic impurities, including catalyst residues, products of oxidation and dissociable end groups. Although low conductivity is the norm, special high-conductivity polymers, based on certain organic molecular structures which are known to exhibit electronic conductivity, have been made. The enthusiasm for the search and discovery of the underlying organic structures which conduct electronically owes much to the biologist Szent-Györgyi (1941) who speculated about the role of semiconduction in living systems. Interestingly enough, semiconduction eventually failed, as a general foundation, to explain complex electron transfer in living cells. Nevertheless, in the meantime, chemists had been led by this work to conjecture that in certain organic compounds, where chemical reactivity had already shown that electrons are extensively delocalised in the molecules, there might be enhanced conductivity, and in 1948 Eley demonstrated conduction in one such case, the phthalo-cyanines.

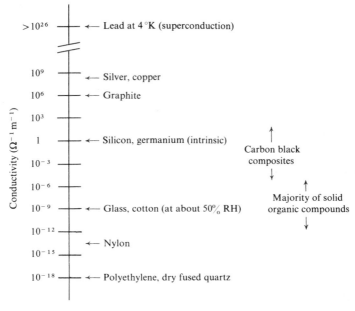

Fig. 5.1. Chart of typical conductivities.

In spite of the great advances that have been made in recent years in developing organic molecular systems which have high electronic conductivity, there is still no inherently conductive polymer which also exhibits the necessary mechanical properties to make a truly conductive plastic material. To some extent, as we shall see, the two sets of properties are mutually exclusive. The major practical solution to the problem where only a modest level of conductivity is required is to make a conducting composite material by dispersing a conductive element, like a metal or carbon black, in an insulating polymer in some way.

We can see that ionic conduction, electronic conduction and conduction in heterogeneous composites are all important in their own way in polymers, but they form three quite distinct subjects and they will therefore be treated separately in the rest of this chapter.

5.2   Ionic conduction

The most definitive evidence for ionic conduction is the detection of electrolysis products formed on discharge of the ions as they arrive at the electrodes. Unfortunately, the very low level of conductivity in polymers generally precludes such detection. Even at a conductivity of $10^{-9}\,\Omega^{-1}\,m^{-1}$, and we must bear in mind that many polymers exhibit conductivities several orders of magnitude lower than this, 100 V applied across a specimen 100 mm^2 in area and 1 mm thick would only produce about 10^{-11} m^3 of gas at NTP per hour. We therefore have to rely on rather more indirect means of elucidating the mechanism of conduction.

When the conductivity in question is very low, there are other experimental difficulties too. On application of a step voltage across a specimen, the initial current may be dominated by a displacement current due to polarisation of the material. Since some dipole orientation may be very slow to reach equilibrium, the displacement current can swamp a small conduction current for a long time. Taking into account the drift characteristics of any practical measurement system, we can appreciate that in extreme cases a steady current reading will never be obtained, and a direct measurement of conductivity will not be possible.

Some physical insight into what a conductivity of around $10^{-9}\,\Omega^{-1}\,m^{-1}$ or less entails may be gauged through equation (5.1). We can tentatively use a mobility value of the order of 10^{-9} m^2 V^{-1} s^{-1}, as found for ionic carriers in hydrocarbon liquids at room temperature. Although we would expect the true value for solid polymers to be lower, it should not differ by more than a factor of 100 or so for small ions. Then, assuming that the ions carry a single electronic charge, substitution in equation (5.1) gives a carrier concentration of 10^{19} m^{-3}. Comparing this with the typical number density of molecular repeat units (nominal molecular weight $= 100$) of about 10^{28} m^{-3}, we can see that an exceedingly low carrier concentration is implied. This means that degrees of ionic impurities which may be totally ignored in the context of other properties may have a significant effect on conductivity.

In some instances evidence favouring ionic conduction is provided by a strong correlation between dielectric constant and conductivity, which is readily explained by the reduction of the Coulombic forces between ions in a high dielectric constant medium. This renders the dissociation energy of an ionic compound inversely proportional to the static dielectric constant ε_s. Consider the dissociation reaction:

$$AB \rightleftharpoons A^+ + B^-,$$
$$(1-f)n_0 \quad fn_0 \quad fn_0 \tag{5.2}$$

where the original concentration of the ionic compound is n_0 and its fractional degree of dissociation is f at equilibrium. Applying the law of Mass Action, we may then define an equilibrium constant K in terms of the concentrations of the reactants and resultants, as follows:

$$K = \frac{[A^+][B^-]}{[AB]} = \frac{f^2 n_0}{1-f}. \tag{5.3}$$

The equilibrium will be governed by the change in free energy ΔG for the reaction, so that

$$K \propto \exp\left(-\frac{\Delta G}{kT}\right) = K_0 \exp\left(-\frac{\Delta W}{\varepsilon_s kT}\right) \tag{5.4}$$

where ΔW is the energy required to separate the ions in a medium of unit dielectric constant, and entropy terms are taken into the constant K_0. If AB is the only ionisable species present, the conductivity will be given by

$$\sigma = fn_0 e(\mu_+ + \mu_-) \tag{5.5}$$

where μ_+, μ_- are the mobilities of the positive and negative ions, respectively, and e is the magnitude of the charge on an electron. For a small degree of dissociation, equation (5.3) becomes

$$f \approx \left(\frac{K}{n_0}\right)^{\frac{1}{2}}, \tag{5.6}$$

and substitution from equations (5.4) and (5.6) in (5.5) gives:

$$\sigma = (K_0 n_0)^{\frac{1}{2}} e(\mu_+ + \mu_-) \exp\left(-\frac{\Delta W}{2\varepsilon_s kT}\right). \tag{5.7}$$

The presence of the dielectric constant in the exponent of equation (5.7) means that it will exert a strong influence on conductivity. In this way the absorption of water, which has a relatively high dielectric constant, generally enhances the conductivity of a polymer greatly, and polymer–water systems frequently conform to the equation

$$\log \sigma = A\varepsilon_s + B, \tag{5.8}$$

where A and B are constants.

Equation (5.7) also shows that (*a*) $\sigma \propto n_0^{\frac{1}{2}}$, and (*b*) the apparent activation energy obtained from the slope of the Arrhenius plot of log σ against $1/T$ is $\Delta W/2\varepsilon_s$. The square-root dependence of σ on n_0 and the occurrence of the factor $\frac{1}{2}$ in the activation energy both stem from the control of the ionic dissociation equilibrium by the law of Mass Action.

Further evidence of ionic conduction may be obtained from studies of the dependence of current on applied voltage for which a theoretical expression may be derived on the basis of a simple model (Mott and Gurney, 1948). Suppose that the unit motion of an ion in the absence of a field is a jump within the matrix of polymer molecules to a neighbouring position of exactly equal energy, passing over a potential-energy barrier of height ΔU^*. The ion will be in a constant state of vibration (frequency v) when lodged in a potential well, and we can assume that the probability that it will pass over the barrier is $e^{-\Delta U^*/kT}$ in each vibration, or $v e^{-\Delta U^*/kT}$ per second. Now consider the effect of applying a uniform electric field E. In the direction perpendicular to the field the potential-energy barrier will be unaffected, but in the direction of the field, and against the field, the barrier heights will be changed by $\mp \frac{1}{2}eEa$, respectively, where a is the distance between neighbouring potential wells (see fig. 5.2). The probability that the ion will move in the direction of the field will now be

$$v e^{-(\Delta U^* - \frac{1}{2}eaE)/kT},$$

and in the opposite direction

$$v e^{-(\Delta U^* + \frac{1}{2}eaE)/kT}.$$

The mean drift velocity u in the field will therefore be

$$u = v e^{-\Delta U^*/kT} 2 \sinh (eaE/2kT). \tag{5.9}$$

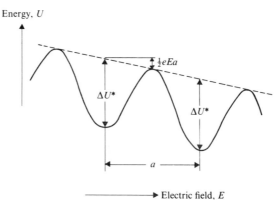

Fig. 5.2. Diagram showing the deformation of ionic potential-energy wells by an applied electric field.

Taking into account that the concentration of ions is proportional to $e^{-\Delta W/2\varepsilon_s kT}$, the current density j flowing through a specimen across which an electric field E is applied will be given by

$$j \propto e^{-(\Delta W/2\varepsilon_s + \Delta U)/kT} \sinh (eaE/2kT), \tag{5.10}$$

or, more simply, at constant temperature,

$$j \propto \sinh (eaE/2kT). \tag{5.11}$$

The graph in fig. 5.3 shows how well the results (Kosaki, Sugiyama and Ieda, 1971) for a sample of PVC fit this formula over a wide range of

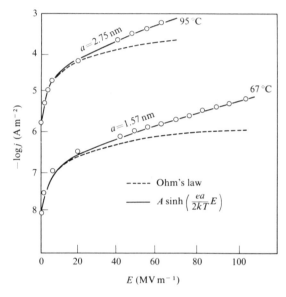

Fig. 5.3. Graphs showing the effect of electric field strength on conduction in poly(vinyl chloride) (Kosaki, Sugiyama and Ieda, 1971).

fields. From this curve one can also derive a value for the jump distance a. It is found that below T_g (87 °C for the PVC specimen) the apparent jump distance is about 1.2 nm, a value which is quite consistent with typical molecular spacings.

Although it is not possible to identify the ions experimentally, we may reasonably assume that they are mainly derived from fragments of polymerisation catalyst, degradation and dissociation products of the polymer itself, and absorbed water. On this basis a polymer like PVC most probably contains H_3O^+, Na^+, K^+ cations and OH^-, Cl^-, Br^- anions.

A feature which is common to most polymers is a change in temperature dependence of conduction as the temperature is raised through the glass transition region. Now we may expect that free volume, which affects the rate of molecular motions at the glass transition (see §3.3), will be equally important to the mobility of ions. Miyamoto and Shibayama (1973) have taken this into account by assuming that the probability per second of a unit ion jump in the absence of an electric field is proportional to $vP_v e^{-\Delta U^*/kT}$, where P_v is the probability that a hole of sufficient size is available for the ion to jump into, i.e.

$$P_v = e^{-\gamma v_i^*/v_f}, \tag{5.12}$$

where v_i^* is the critical free volume for an ion jump, v_f is the average free volume per molecular segment, and γ is a numerical factor to correct for the overlap of free volume. The current density equation (5.10) may then be modified to become

$$j \propto \exp\left(-\frac{\gamma v_i^*}{v_f} - \frac{\Delta W/2\varepsilon_s + \Delta U}{kT}\right) \sinh\left(eaE/2kT\right). \tag{5.13}$$

Taking free volume characteristics from WLF data, a straight line is obtained when $\log \sigma + \gamma v_i^*/2.303 v_f$ is plotted against $1/T$ for polystyrene through the glass transition region as shown in fig. 5.4, in accord with the behaviour expected from equation (5.13).

The greater ease with which ions may be reasonably expected to move

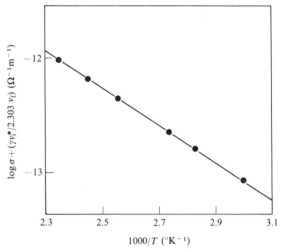

Fig. 5.4. The temperature dependence of the conductivity of polystyrene through the glass transition region (Miyamoto and Shibayama, 1973).

through amorphous regions, as compared with crystalline regions, ties in with the general fall in conductivity with crystallinity observed in most polymers.

Electrostatic problems on insulating plastics (see chapter 7) are often alleviated by the use of so-called antistatic agents. These are usually surface-active chemicals which form a thin (10 nm or less) layer on the outer surface of the plastic moulding, film etc. In most cases their operation additionally requires the uptake of water in the surface layer to promote ionic dissociation and to provide a medium for ionic flow. Dissipation of charges by conduction through the surface layer of such systems therefore works best in moist atmospheres. Although performance may be satisfactory at normal ambient relative humidities of 50% or so, conduction may fail in very dry conditions, depending on the exact nature of the surface layer. Sometimes antistatic agents are applied directly to the surface by a dip or spray method. Alternatively, they are added (at about the 0.1% level) to the bulk of the polymer powder before processing. Although this is a very convenient process, the effect takes time to develop, because enough antistatic agent must first diffuse outwards from the bulk to produce the conductive *bloom* on the final surface. Improved, more permanent and durable protection against static charges can sometimes be obtained by the incorporation of an ionic copolymer.

Our considerations above have laid emphasis on conduction by ions present by default or added as a separate ingredient to the polymer. Some polymers, however, are themselves inherently conductive in having at least one ionisable group per monomer unit, the most notable being cellulose and its derivatives. Polyamides also show pronounced ionic conduction effects at elevated temperatures (the conductivity of Nylon-6,6 intrinsically exceeds $10^{-8}\,\Omega^{-1}\,m^{-1}$ above 100 °C), evidently as a result of the dissociation of amide groups to give protons. There has been much discussion about the possibility of a particularly fast mechanism for the transfer of these protons through a lattice of hydrogen-bonded polyamide molecules, analogous to the conduction process in ice, but the case is difficult to prove unambiguously.

5.3   Electronic conduction

Electronic conduction in organic, molecular compounds differs in several important ways from the more familiar kind in metals and inorganic semiconductors like silicon and germanium. That is not to say that they are separate subjects and, indeed, the well-known band theory of atomic lattices has provided the essential basis of concepts and language for the discussion of conduction in molecular solids. A detailed treatment of band theory would be out of place here, but we include the following descriptive review of those elements which are of most relevance to the development of our subject.

5.3.1   **Band theory**

When two hydrogen atoms come together so that their 1s-electron orbitals overlap, two new σ electronic orbitals are formed around the atoms, symmetric with respect to the interatomic axis. In one orbital, the bonding orbital, an electron has a lower energy than in the isolated atom orbital, and in the other, the antibonding orbital, an electron has a higher energy (fig. 5.5). The two electrons from the hydrogen atoms, if they are of opposite spin, may *pair* in the bonding orbital to give a stable molecule, with total energy less than the sum of two isolated hydrogen atom energies. In an extended, crystalline solid like silicon, where many atoms strongly interact, similar splittings of energy levels also occur. Sets of energy levels this time form two continuous energy bands, called the valence band and the conduction band, analogous to the bonding and antibonding levels of the two-atom molecule, the energy gap between them representing a forbidden zone for electrons.

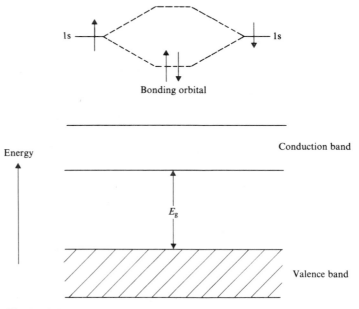

Fig. 5.5. Schematic interaction of hydrogen atoms and energy bands.

An important feature of the band system is that electrons are delocalised or spread over the lattice. Some delocalisation is naturally expected when an atomic orbital of any atom overlaps appreciably with those of more than one of its neighbours, but delocalisation reaches an extreme form in the case of a regular, 3-dimensional lattice. We can understand this best if we choose to think of the wave nature of electrons, and from that point of view we can formulate band theory as follows.

A free electron propagates through space as a wave* characterised by a wavevector $k(k = 2\pi/\lambda$ where λ is the wavelength), whose magnitude is related to the momentum p of the electron by the fundamental relation

$$k = 2\pi p/h, \tag{5.14}$$

h being Planck's constant. In a lattice of strongly interacting atoms where the orbitals of the valence electrons overlap well, the valence electrons are *nearly* free to roam anywhere within the confines of the lattice. (The inner cores of electrons remain, of course, localised on the atoms.) The regular array of atomic centres does impose a certain restriction on the valence electrons: in the same way that X-rays suffer Bragg reflections by the lattices of crystals (the basis of X-ray crystallography), electron waves which satisfy the Bragg reflection condition will be unable to pass through the lattice. For a simple linear lattice with a spacing a, the Bragg condition is

$$k = \pm n\pi/a, \tag{5.15}$$

where n is an integer. The excluded waves correspond to gaps in the allowed values of electron energy, and the theoretical result for a simplified model of a linear monatomic array is shown in fig. 5.6, set alongside that for the completely free electron case. The situation is similar for 3-dimensional lattices, the allowed values of the wavevector forming regions, called Brillouin zones, in k-space separated by forbidden energy gaps.

For a simple linear lattice of length $L = Na$, where N is the total number of atoms, it may be shown that the total number of electron wave states in the lowest energy band is exactly N. This means that each cell of the lattice contributes one independent k-value to the energy band. This result applies to every energy band of the system and also carries over to 3-dimensional lattices. Allowing for two spin orientations of an electron, the Pauli exclusion principle then tells us that there will be room for two electrons per cell in an energy band. Hence, if each atom

* Free electron wave functions are of the form:
   $$\psi_k(r) = (1/V)^{\frac{1}{2}} e^{i k \cdot r},$$
where r is the position vector and V is the volume of the containing space.

contributes one bonding electron, the valence band will be only half-filled. Taking into account the full detail of actual valence orbitals of component atoms, valence bands become a good deal more complicated than indicated above. For instance, it is common for different sets of valence electrons to interact and form composite bands.

In a full band there can be no net flow of electronic charge under an external electric field. The reason is that for every wave state in which an electron is travelling in one direction there is another in which the electron is travelling in the opposite direction and there are no spare states (equal numbers of $+k$ and $-k$ values). This situation applies to

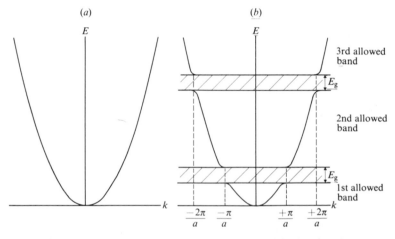

Fig. 5.6. Relation between energy E and wavevector k for (a) a free electron, and (b) a *nearly* free electron in a monatomic linear lattice (spacing a).

pure silicon in its ground state at the absolute zero of temperature, where the bonding electrons provided by each silicon atom just fill the valence band. *Intrinsic* conduction can therefore only occur when electrons are promoted across the band gap into the conduction band by some means. Then both this electron and the hole (a vacancy with an effective positive charge) in the valence band can contribute to net charge flow. Promotional energy can be obtained by direct photon absorption, or by raising the temperature when an occasional electron may receive sufficient thermal energy from the lattice (electron–phonon collision).

Metals have partially filled valence bands or overlapping bands and therefore conduct well without the need for promotion of electrons across an energy gap. As electrons obey Fermi–Dirac statistics, the distribution function giving the occupation probability $f(E)$ of a state of energy E at equilibrium is

$$f(E) = \frac{1}{1 + e^{(E - E_F)/kT}}. \tag{5.16}$$

The parameter E_F, called the Fermi energy, can be regarded as the chemical potential of the electrons and corresponds to an energy level where the probability of a state being occupied is $\frac{1}{2}$. At the absolute zero of temperature all electrons fall into the set of states with the lowest energy and the Fermi energy of a metal then becomes identical with the energy of the highest occupied state in the partially filled valence band. At higher temperatures some states above E_F become filled at the expense of states below E_F.

Detailed statistical analysis (e.g. see Kittel, 1966) shows that for intrinsic band conduction in a semiconductor, where the concentration of conduction electrons n always equals the concentration of holes p, and where the energy band gap E_g is wide (i.e. $E_g \gg kT$),

$$n = p = 2\left(\frac{2\pi m^* kT}{h^2}\right)^{\frac{3}{2}} e^{-E_g/2kT}, \tag{5.17}$$

where m^* is the effective mass of the electron or hole. Thus, as the temperature is increased, the charge-carrier concentration increases strongly with temperature. This dominates the temperature dependence of the conductivity, giving it an Arrhenius-like character with an effective activation energy of $\frac{1}{2}E_g$. The factor $\frac{1}{2}$ in the activation energy reflects the dynamic equilibrium of electron and hole concentrations by continual thermal generation and recombination in accord with the law of Mass Action (compare with equation (5.7)).

One other feature of the energy scheme which is especially important in semiconductors is the occurrence of levels in the normally forbidden energy zone due to the presence of impurities. Deliberate doping with elements which give donor states near to the conduction-band edge and electron acceptor states near to the valence-band edge can produce enormous increases in the population of conduction electrons or holes at a given temperature. This is called *extrinsic* conduction.

In metals, where the valence bands are only partially filled even in the ground state, the effective concentration of carriers is scarcely affected by temperature. Actually, the conductivity *decreases* slightly with increasing temperature, because at higher temperatures the lattice vibrations scatter electrons and their *mobility* goes down. For the same reason mobility also decreases with temperature in the case of silicon, but the conductivity still increases with temperature, because the steep rise of carrier concentration swamps the more gradual decline in mobility.

The width of an energy band reflects the strength of the interactions amongst a particular set of atomic orbitals, the interaction depending on the degree of overlap of the orbitals. For large interatomic separation, the overlap is necessarily poor, the band is narrow and, as one may expect, now that the perturbations from the lattice vibrations are relatively large, the carrier mobility is low. It may be shown that when the overlap is so poor that the mobility falls below 10^{-4} m^2 V^{-1} s^{-1}, it is no longer meaningful to use band theory – the mean free path of an electron between consecutive scattering collisions is less than both the free-electron wavelength and the interatomic spacing of the lattice.

5.3.2   Hopping conduction

Disorder in a lattice affects both the energetic and spatial distribution of electronic states. For a random distribution of atoms the density of

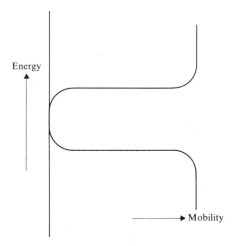

Fig. 5.7. A mobility gap.

electronic energy states *tails* into what is normally the forbidden zone, and the electrons in these tails are localised. There is then not so much an energy gap as a mobility gap (see fig. 5.7). In other words there is an intermediate range of electronic energy states in which mobilities are very low. Only when electrons are excited to higher energy states, in which mobilities are higher, can appreciable conduction occur.

Conduction via localised electrons implies discrete jumps across an energy barrier from one site to the next (fig. 5.8). An electron may either

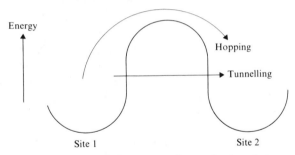

Fig. 5.8. Diagram of electron-transfer mechanisms between adjacent sites separated by a potential-energy barrier.

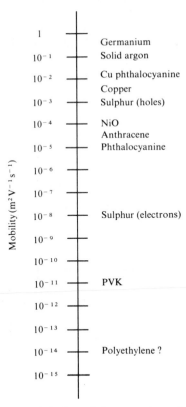

Fig. 5.9. A chart of charge-carrier mobilities.

hop over, or tunnel through, the top of the barrier, the relative importance of these two mechanisms depending on the shape of the barrier and the availability of thermal energy. This thermally activated type of mobility will increase with temperature, of course, in contrast with that in band conduction. It is apparent that we learn a lot about the conduction process by determination of carrier mobility, e.g. its temperature dependence provides a good criterion by which we can distinguish band and hopping types of mechanism. Data for a selection of systems are displayed on the mobility chart (fig. 5.9).

5.3.3   Organic molecular solids and polymers

Let us now turn to organic molecular solids. The essential new feature here is that groups of atoms are chemically bonded together in discrete molecules, which in turn are held together by relatively weak van der Waals forces. We must therefore distinguish *intra-* and *inter*molecular types of electronic motion. We can immediately foresee that transfer of electrons from one molecule to another may be the main stumbling block for good conduction on the macrosopic scale. Some simple molecular compounds whose molecules are of a suitable shape for close-packing in crystals show electronic mobilities which just come within the bounds of band theory. Crystalline anthracene, for example, which consists of flat molecules, shows a mobility of about 10^{-4} m^2 V^{-1} s^{-1}, which decreases with increase in temperature. We can therefore say that in favourable cases a molecular conduction band exists. In other cases hopping conduction occurs. Thus Gibbons and Spear (1966) clearly observed hopping conduction in sulphur crystals (arrays of crown-shaped S_8 rings) with a hopping activation energy of about 0.25 eV.

In a system of long polymeric molecules one could assert with some justification that the intermolecular conduction problem is much less important, because relatively few intermolecular transfers are required. So we must now ask what are the prospects for high intramolecular conduction.

If we regard each molecule for the moment as a miniature lattice we shall have:

(*a*) A definitely spaced series of atoms with rigidly fixed distances between nearest neighbours. (Unless the molecules are in a crystalline phase considerable twisting about bonds may be present.)

(*b*) Small separation of atoms giving good overlap of atomic orbitals.

(*c*) A full valence shell, analogous to a full valence band.

(*d*) A very large excitation energy to the lowest excited electronic state, corresponding to strong chemical binding, i.e. a wide band gap like that in diamond.

On this basis, McCubbin and Gurney (1965) applied band theory to a single, long polyethylene molecule in its planar zigzag conformation found in the crystalline phase:

$$\text{–CH}_2 \diagup \text{CH}_2 \diagdown \text{CH}_2 \diagup \text{CH}_2 \text{–}$$

Although such calculations predict a very wide band gap (greater than 5 eV) corresponding to the strong chemical bonds between carbon atoms in the polymeric molecule, the estimated carrier mobility is about 5×10^{-3} m^2 V^{-1} s^{-1} (for holes), which is comparable to that in metals. Now we may anticipate that experimental detection of this particular type of electronic conduction would be difficult, partly because such high-energy electrons are involved, and partly because a high degree of chain regularity is required to preserve the band structure and avoid trap formation. In fact, no unambiguous observation of it has yet been achieved, although excited-state energies of the right order of magnitude have been found by scattered-electron-spectroscopy techniques (Delhalle *et al.*, 1974).

5.3.4   Conjugated chains

We can understand, then, that long polymeric molecules which are fully saturated in the chemical sense (no double bonds) will not have any significant electronic conductivity under normal circumstances. The situation is different when we consider the possibility of chemical unsaturation. If each carbon atom along a chain has only one other atom, e.g. hydrogen, attached to it, the spare electron in a p_z-orbital of each carbon atom overlaps with those of carbon atoms on both sides, forming delocalised molecular orbitals of π-symmetry (fig. 5.10). In chemical terms this is a conjugated chain and may be represented as a sequence of alternating single and double bonds:

In a benzene ring the π-electrons are delocalised over the whole sequence of six carbon atoms in this way, and induced ring currents, which are responsible for anomalous magnetic susceptibilities in aromatic compounds, are evidence of this.

If π-orbital overlap extends throughout a very long chain, the discrete set of molecular electronic states may be expected to merge into a half-full valence band, giving metallic-like conduction within the molecular chain. When a conjugated chain contains a finite (and even) number of carbon atoms, the electrons in the ground state will be paired in the lowest energy set of molecular orbitals. Conduction then requires the excitation of an electron from the highest occupied orbital to the lowest unoccupied one. Now the energy levels for a finite conjugated chain may be calculated by elementary quantum mechanics using the very simple 1-dimensional particle-in-a-box model. This model corresponds quite well to the fairly uniform overlap of p_z-orbitals along the chain. On this basis the excitation energy ΔE from the ground state to

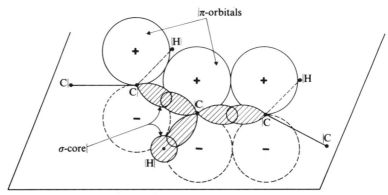

Fig. 5.10. Schematic representation of the bonding electronic orbitals of a conjugated carbon chain.

the first excited state for a linear chain of N atoms, each spaced a distance 0.14 nm apart, is given by the equation

$$\Delta E = 19(N+1)/N^2. \tag{5.18}$$

The important feature is that ΔE is small for a long molecule, and when the chain is of the length typical of polymers ($N \approx 1000$), it becomes comparable with average thermal energies at room temperature (0.025 eV) and a high equilibrium concentration of carriers is therefore predicted. Since the π-orbitals involve good overlap of component atomic p_z-orbitals, the mobility along the chain will be high. Combining these estimates of high carrier concentration with high mobility from this simple model, we predict at least a high intramolecular conductivity and it would not be unreasonable to expect a modest macroscopic conductivity in, say, the polyacetylene family.

   In reality it does not turn out so well in these linear conjugated chains, and only highly crystalline, stereoregular polyacetylenes show appreciable conductivity. The best result so far (10^{-1} Ω^{-1} m^{-1}) has been obtained by Shirakawa and Ikeda (1971) with a thin film of *trans*-polyacetylene. In most conjugated polymers the conductivity is little better than that in ordinary polymers. Apart from the intermolecular conduction difficulty, there are two major reasons why this should be so.

First, in long chains *alternation* tends to set in. Instead of equal bond lengths along the chain, alternately long and short bonds occur, and delocalisation becomes self-limited. In other words, as the chain is increased in length, the π-electrons are not delocalised over a correspondingly greater distance, and the activation energy ΔE for generation of carriers no longer falls linearly with chain length. The alternation (an example of Jahn–Teller stabilisation) is predicted by more elaborate quantum-mechanical calculations and can be observed in the absorption spectrum of such molecules.

Secondly, it is clear that for conjugation to be fully maintained the molecules must remain planar. When rotation about any of the bonds takes place, conjugation is *broken,* because the atomic p-orbitals no longer overlap properly. Polyphenylacetylene provides a typical example of this kind of failure of conjugation; steric hindrance from the phenyl side groups is sufficient to prevent the chain from taking up a planar conformation over any significant length. The conductivity is below $10^{-12}\,\Omega^{-1}\,m^{-1}$ at room temperature. Even in an unhindered case, entropy considerations strongly favour out-of-plane conformations for long chains.

We conclude that the full potential for high conductivity in conjugated chains will rarely be realised, because only in exceptional circumstances will an undistorted structure be maintained. In one very notable example, poly(sulphur nitride), metallic conductivity was nevertheless discovered (Greene, Street and Suter, 1975). This unsaturated polymer $(SN)_n$ crystallises in a fibrillar form with its molecules extended along the fibre axis in helical conformation. The conductivity, which is highly anisotropic, may be as high as $10^5\,\Omega^{-1}\,m^{-1}$ in the direction of the fibre axis. The structure appears to survive distortion to very low temperatures, and the solid even exhibits superconductivity (see later) below 0.3 °K. Weak interactions between the chains, producing some 3-dimensional band structure, may be the cause of this.

One way round the difficulties of achieving conductivity in conjugated chains is to lock the molecules into the required configuration, as in a ladder-type polymer:

A whole collection of polymers called poly(acene quinone radical) (PAQR) polymers with structures of the type and of the order of 500 nm in length, have been made (Pohl and Engelhardt, 1962) and these are indeed found to have high conductivity, in the range 1 to $10^{-7} \, \Omega^{-1} \, m^{-1}$. Unfortunately, the high conductivity is achieved at the expense of mechanical properties; the inflexible long molecules make such an intractable material – infusible and insoluble in all solvents – that even characterisation is difficult by conventional techniques. Apart from a high overall conductivity it is interesting to note, however, that such compounds show extraordinarily high dielectric constants in the range 20 to 50 000 (Hartman and Pohl, 1968). This is explained in the following way.

The existence of highly extended π-orbitals along the chains means that excitation energies for the transfer of an electron from one molecule to another are low, so that there will be a high equilibrium concentration of radical ions, i.e. chains with an extra electron (or electron vacancy) together with an unpaired spin. (The presence of unpaired spins is confirmed by electron-spin-resonance spectroscopy.) Under the influence of an electric field these electrons or vacancies will behave as efficient intramolecular carriers giving rise to very large electronic displacements of 100 nm or more. This polarisation reaches saturation at quite weak fields, because the carriers need little force to send them to the ends of the chains.

Ladder polymers may be extended even more to give, ultimately, large sheets of fused rings like those in graphite. Here the π-electrons are completely delocalised in two dimensions and metallic-like conduction is found, up to $10^7 \, \Omega^{-1} \, m^{-1}$.

Graphitic-like structures are readily formed by pyrolysis of many organic polymers. The product, of course, is not easy to fabricate into a desired form and it is therefore not very useful. One solution is to pyrolyse *after* fabrication. Thus drawn fibres of polyacrylonitrile readily pyrolyse to give conducting filaments called *black Orlon*:

This material still contains nitrogen and the conductivity is still only $10^{-1} \, \Omega^{-1} \, m^{-1}$. The more recent development of this theme takes the pyrolysis a stage further to give high-tensile carbon fibres with conductivities of the order of $10^5 \, \Omega^{-1} \, m^{-1}$. In this latter process the nitrogen is eliminated altogether, and considerable reorganisation of atoms, helped by the original orientation of the molecules in the drawn fibre, takes place to give a graphitic structure.

5.3.5   Charge-transfer complexes

There are several rather special classes of organic compound which show high electronic conductivity. The most notable are charge-transfer complexes and the closely related radical ion systems. These may be incorporated into polymeric chains whilst still retaining to a large extent their conductive properties, although very few industrially useful materials have so far emerged.

Charge-transfer complexes are formed by partial transfer of an electron from a donor molecule of low ionisation potential to an acceptor molecule of high electron affinity:

$$D + A \rightarrow D^{\delta+} A^{\delta-}.$$

Complete charge transfer would imply the formation of radical ions, that is ions carrying an odd number of electrons. A good example is the reaction between the aromatic hydrocarbon pyrene (conductivity $10^{-12} \, \Omega^{-1} \, m^{-1}$) and iodine (conductivity $10^{-7} \Omega^{-1} m^{-1}$) to give a complex with a conductivity $1.3 \, \Omega^{-1} m^{-1}$:

Charge-transfer complexes pack closely in their crystalline phase through formation of rigid multi-sandwich stacks.

$$\ldots D \ A \ D \ A \ D \ A \ldots,$$

making for a rather brittle solid. The conductivity is highly anisotropic, being greatest along the direction of the stacks, and supposedly occurs by radical-ion disproportionation (Eley, 1967):

$$
\begin{array}{ccccc}
A^{\bar{}} & A^{=} & A^{\bar{}} & A^{\bar{}} & A^{\bar{}} \\
D^{+} & D^{+} & D & D^{+} & D^{+} \\
A^{\bar{}} & A^{\bar{}} & A^{\bar{}} & A^{=} & A^{\bar{}} \\
D^{+} & D^{+} & D^{+} & D^{+} & D
\end{array}
\quad \text{etc.}
$$

In a sense one could say that incipient decomposition and ease of electron transfer are dual aspects of the very nature of charge-transfer complexes.

Many polymeric versions, usually with the donor species attached to the polymer backbone, have been synthesised:

The conductivity is generally lower than that of the corresponding monomeric complex and this may be attributed (Litt and Summers, 1973) to incorrect spacing of repeat units. By selecting polyethylene-imine as the backbone, the spacing of methylmercaptoanisole donor units (0.635 nm) in the crystalline polymer is the same as the repeat distance of most charge-transfer complexes (0.64 to 0.68 nm). If 2,4,5,7-tetrafluorenone is then used as acceptor, the polymeric complex

has a conductivity 10^3 times that of the monomeric one. This vividly illustrates the way in which molecular geometry can be tailored to optimise the conductive property.

Even polymeric charge-transfer complexes are still quite brittle materials and so they can only be used where mechanical properties are not important. Nevertheless, one such material, poly(2-vinyl pyridine)–iodine, has found commercial use in highly efficient solid-state batteries. The complex primarily serves as a convenient source of iodine for the cell reaction:

$$2\,Li + I_2 \rightarrow 2\,LiI$$

Being at the same time conductive, it may completely surround the cathode without increasing the internal resistance of the cell too much.

5.3.6   Radical-ion compounds

Central to the development of highly conductive material based on radical-ion systems has been the discovery of the compound 7,7,8,8-tetracyanoquinodimethane (TCNQ) by Du Pont workers (Acker *et al.*, 1960). This molecule is a very strong electron acceptor, forming first the radical anion and then the dianion:

$$E_1 \qquad\qquad E_2$$

Oxidised   Semireduced   Reduced

In electrochemical terms the interconversions of this system of neutral molecule, radical anion and dianion are characterised by redox potentials $E_1 = 0.127$ V and $E_2 = -0.219$ V, which measure the energy differences between the oxidation states. The most unusual stability of the semireduced radical ion with respect to the neutral molecule mainly accrues from the concomitant change from the relatively unstable quinonoid structure to the aromatic one, allowing extensive delocalisation of the π-electrons over the carbon skeleton. As a consequence of this stabilisation, the TCNQ not only forms typical charge-transfer complexes but is also able to form true radical-ion salts, incurring complete 1-electron transfer. Thus, on the addition of lithium iodide to a solution of TCNQ, the simple lithium TCNQ salt is formed:

$$TCNQ + LiI \rightleftharpoons Li^+TCNQ^- + \tfrac{1}{2}I_2\downarrow.$$

If the free iodine precipitate is removed, the TCNQ salt may be crystallised from the mother liquor. The crystals show remarkably high electronic conductivity of about 10^{-3} Ω^{-1} m^{-1}. Furthermore, the lithium ions may be replaced by certain organic cations to give salts with even higher conductivities, e.g. crystals of *N*-methyl phenazinium TCNQ,

show a conductivity as high as 10^4 Ω^{-1} m^{-1} in certain directions. With some cations TCNQ will form both simple and complex salts, the latter containing an extra equivalent of neutral TCNQ. In these cases, the complex salt generally has the higher conductivity, as exemplified by the quinolinium salts:

X-ray crystallography of the radical-ion salts shows that the TCNQ units, which are planar, are arranged in face-to-face staggered stacks, the associated cations being externally located with respect to these stacks. Conductivity is very anisotropic, with the high conductivity corresponding to movement of carriers along the stack direction. We should note that the situation is in sharp contrast to that in normal charge-transfer complexes where highest conductivity occurs within the alternating stacks of donor/acceptor species.

TCNQ radical-ion salts themselves span a wide range in conductivity, reflecting a high degree of sensitivity of the conduction process along the stacks to the nature of the counter ions. The overlap of the π-orbitals, containing the odd electrons, in a face-to-face stack of TCNQ radical ions is rather like that of the valence orbitals in a row of metal atoms, and we might therefore anticipate 1-dimensional metallic-like conduction with a half-full valence band. The chief difference is that the positive counter ions, unlike the positive atomic cores in a metal, are laterally displaced from the system of overlapping electronic orbitals, rendering the overall binding not so tight, and electron–electron repulsive forces relatively more important. The effect may be partially offset by using an aromatic counter ion with a large electronic polarisability. This helps to reduce repulsion between any two electrons: crudely, one electron induces a dipole in the cation, this dipole then interacting attractively with the second electron. Thus in the N-methyl phenazinium salt,

the interplanar spacing is 0.324 nm (less than the distance between adjacent molecular sheets of graphite) and metallic band-type conduction obtains. It is interesting that although the conductivity of such metallic-like salts at first increases as the temperature is lowered (a characteristic feature of metallic conduction), the conductivity eventually peaks, falling to low values again at very low temperatures. This is taken as evidence of a metal–insulator transition of the kind that Peierls (1955) predicted for a 1-dimensional solid. The effect is attributed to a distortion, analogous to the Jahn–Teller alternation in a conjugated chain, which gives a more stable structure below a certain temperature. In effect, the unit cell doubles in size, thereby halving the number of electronic states in the first Brillouin zone. There are then no empty states in the valence band and states below the band gap are no longer *wasted* energywise. The electron energy is thereby minimised, whilst eliminating conduction without activation of electrons across the band gap into the conduction band (second Brillouin zone). The saving in electron energy must be set against the increase in elastic energy due to the distortion of the molecular lattice; the balance between the two factors determines the temperature of transition.

Unsuitably shaped counter ions easily upset the metallic-like array of TCNQ radical ions. Typically the simple salts become arranged as dimers with enlarged interplanar spacings of 0.332 nm between dimers, and conduction reverts to a hopping-type. In the case of N-ethyl phenazinium TCNQ, the conductivity is eleven orders of magnitude less than that of the N-methyl derivative! In such cases the difficult step in the conduction process is the generation of a carrier, which entails overcoming large Coulombic forces in forming the dianion:

Carrier generation

Transport

In complex salts on the other hand, the dianion need not be involved at all, and this explains the generally superior conductivity of the complex salts:

Carrier generation

Transport

It also explains how doping a simple salt with a small amount of neutral TCNQ can change the conductivity enormously.

Quite soon after the discovery of conduction in TCNQ salts, researchers turned their attention to the synthesis of polymeric versions, their aim clearly being to combine the toughness and processability of a material based on long flexible molecules with the facile electronic transport of a radical-ion salt. Of the many approaches the simplest is probably to use a polymeric cation, the ionic centres ideally being polarisable and aromatic, as a backbone along which to string TCNQ anions. Poly(2-vinyl pyridine)

and its derivatives have often served as a basis for this. One might hope to optimise the conductivity by selecting a particular (stereoregular) polymer that would adopt the right conformation in the solid to allow the associated TCNQ units to form the necessary face-to-face stacks for conduction. However, in spite of considerable preparative effort involving many structural variants, only limited success has been achieved, and the highest conductivities (about $1\ \Omega^{-1}\ m^{-1}$) are much less than those of the corresponding non-polymeric salts. Nonetheless these polymers can be dissolved in organic solvents and cast into films. Such films are black, are somewhat brittle and deteriorate in air, but they probably represent the most dramatic combination of plastics properties and electronic conduction so far devised. Noteworthy elastomeric salts containing 5% by weight TCNQ have also been prepared. Although these have only modest conductivities (around $10^{-6}\ \Omega^{-1}\ m^{-1}$), they retain their elasticity and, on stretching up to 80%, their conductivity is not destroyed.

There are also a few examples of radical-cation systems which exhibit similar conduction phenomena to radical-anion ones. Particularly stable radical cations are based on a category of heterocyclic compounds called violenes, which contain even polyene structures:

$$X\!-\!(CH\!=\!CH\!-\!)_n X \underset{+e}{\overset{-e}{\rightleftharpoons}} \overset{+}{X}\!-\!(CH\!=\!CH\!-\!)_n X \underset{+e}{\overset{-e}{\rightleftharpoons}} \overset{+}{X}\!=\!(CH\!-\!CH\!=\!)_n \overset{+}{X}$$

Combination of one example, tetrathiafulvalene (TTF),

with TCNQ gives a compound which has a very high conductivity (about $10^5 \, \Omega^{-1} \, m^{-1}$) and which apparently contains both radical-cation and radical-anion stacks (charge transfer is about 80% complete).

In a similar way to that in which traces of neutral TCNQ can greatly increase the conductivity of TCNQ simple salts, the presence of a little radical-cation salt can enhance the conductivity of neutral violenes:

Polymers based on a neutral violene doped with a radical-cation salt offer attractive prospects, because such systems are only lightly coloured and are very stable (Goodings, 1976). They are therefore likely to be more useful than TCNQ systems which are so strongly coloured and sensitive to air.

5.3.7   Organometallics

Organometallic groups in polymers may work in several different ways to improve electronic conductivity. The metal d-orbitals may overlap with π-orbitals of the organic structure and thereby extend electron delocalisation along a molecule. Being rather diffuse, the d-orbitals may also serve to bridge molecules in adjacent layers of a crystal. Thus the high conductivity (10^{-3} to $10^{-2} \, \Omega^{-1} \, m^{-1}$) found by Dewar and Talati (1964) in the polymeric square-planar Cu^{II} complex of 1,5-diformyl-2,6-dihydroxynaphthalene dioxime,

may be ascribed to a combination of both of these effects.

When a sequence of complexed transition-metal atoms are present in mixed oxidation states, they may contribute a conductive pathway that is quite separate from the organic framework. The process of electron transport depends on direct shuttling of electrons between metal atoms in different oxidation states, a mechanism that is reminiscent of that in TCNQ complex salts. It is well illustrated by polyferrocenylene,

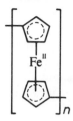

in which Pittman, Sasaki and Mukherjee (1975) have observed a 10^7–10^8-fold increase in conductivity when 35–65% of the ferrocenium (Fe^{II}) centres are oxidised to the ferricenium (Fe^{III}) form with I_3^- as counter anion. In the Krogmann (1969) salts the metal–metal conduction reaches an extreme where chains of metal atoms interact sufficiently strongly to form a conduction band. To understand how this may

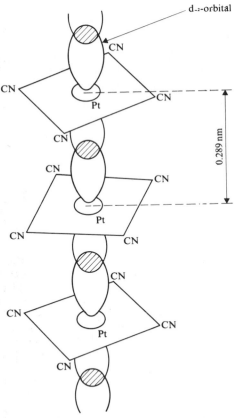

Fig. 5.11. Schematic representation of the bonding arrangement of cyanoplatinate groups in the Krogmann salt $K_2Pt(CN)_4Br_{\frac{1}{3}}$, $2\frac{1}{3}H_2O$.

happen, let us consider the well-known example derived from potassium cyanoplatinate (KCP), as shown in fig. 5.11. In the square-planar complex of KCP the CN^- groups bond with the d_{xy}-orbitals of the Pt atoms, so that the antibonding d_{z^2}-orbitals (perpendicular to the plane of the complex) are fully occupied. Fractional oxidation of KCP with Br_2 removes a third of the antibonding electrons, whereafter the Pt atoms approach closer to each other, and the increased overlap of the d_{z^2}-orbitals then forms a five-sixths-full conduction band. This amounts to a *polymerisation* of the Pt atoms into a 1-dimensional metal. The resulting conductivity is indeed metallic and anisotropic and may be as high as $3 \times 10^4\ \Omega^{-1}\ m^{-1}$ in the direction parallel to the metallic chains. The crystals also exhibit evidence of a metal–insulator transition as the lattice suffers the expected Peierls distortion at low temperatures (about $100\ °K$).

High conductivity in polymeric organometallic complexes is synonymous with very rigid structures giving poor mechanical properties. They are mostly refractory black powders which are insoluble in organic solvents, and show no promise, so far, for practical uses.

5.3.8   Photoconduction

Before discussing photoconduction in polymers we begin this section with a brief, general account of the phenomenon.

Exposure of a semiconductor to light or other electromagnetic radiation may produce a temporary increase in the population of free charge carriers, and the resulting extra flow of current under the influence of an applied electric field is called photoconduction. The photons of the radiation can interact with the semiconductor in a variety of ways to generate carriers. The simplest process is one of absorption of single photons to promote an electron directly from the valence band to the conduction band to give an electron–hole pair, thereby enhancing the concentration of intrinsic carriers. For this to be possible the photon energy must exceed the band gap and there is consequently a threshold wavelength for this type of photoconduction. The threshold is also apparent in the absorption spectrum itself where an absorption *edge* occurs at the frequency corresponding to the onset of electronic transitions across the band gap. Generation of carriers by photon absorption, however, often proceeds in a more indirect manner. For instance, the first step may be the production of excitons, which are localised, but mobile, excited electronic states, and which cannot by themselves transport charge. Two excitons may subsequently collide to produce an electron–hole pair or an exciton may migrate to the surface and react with a surface state to inject a carrier of one sign into the bulk. The latter process is referred to as *photoinjection* of carriers. When the light is switched off, the photoconduction will decay as the carrier population gradually returns to equilibrium. By studying photoconduction kinetics it is often possible to determine the dominant mechanism of carrier loss: neutralisation at electrodes, recombination of electrons with holes, or trapping at defects or impurity centres.

Pulsed photoconductivity provides a powerful means of measuring carrier mobilities. The usual experimental arrangement shown in fig. 5.12 (Kepler, 1960) uses a sandwich cell with a transparent front electrode. If the coefficient of absorption of the light by the photoconductor is high, a flash of light effectively generates a sheet of carriers near to the front electrode. Depending on the direction of the electric field applied across the sample, positive or negative carriers then drift towards the opposite electrode, producing a current in the external circuit; the current pulse terminates as the drifting carriers reach the electrode. The inset in fig. 5.12 shows a typical trace of a current transient obtained on

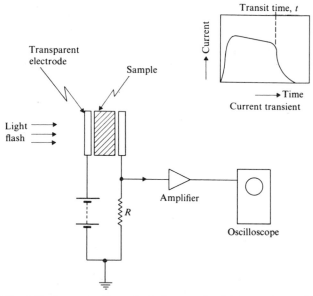

Fig. 5.12. Diagram of a pulsed photoconductivity experiment. The inset diagram shows the form of the oscilloscope trace.

an oscilloscope on which the transit time *t* may be measured. Knowing the thickness of the sample *d* and the applied voltage *V*, the mobility of the carrier is then given by

$$\mu = \frac{d^2}{Vt}. \tag{5.19}$$

In some cases the drift of carriers may be seriously interrupted by capture at trapping sites in the solid. If the traps are energetically shallow, i.e. the depth of the potential well is comparable with thermal energies (kT), the carriers will soon escape again, and the only effect will be to reduce the apparent mobility. The *trap-limited* mobility μ_T will be given by

$$\mu_T = \mu \frac{\tau_c}{\tau_c + \tau_T}, \tag{5.20}$$

where μ is the mobility of carriers between traps, τ_c is the mean lifetime of carriers between traps, and τ_T is the mean time spent in a trap. In this case the apparent mobility will increase with temperature on account of the more rapid thermal release from the traps at higher temperatures:

$$\tau_T \propto e^{U_T/kT}, \tag{5.21}$$

where U_T is the depth in energy of the trap.

Massive photoinjection of carriers into a material where traps greatly reduce the carrier mobility gives rise to a space-charge effect analogous to that in a thermionic valve. As shown by Mott and Gurney (1948) the current density j at low fields, where the space-charge field is important, varies with applied voltage V (sample thickness d) as follows:

$$j \propto \frac{V^2}{d^3}, \tag{5.22}$$

where the current increases as the square of the voltage. In this regime the interpretation of results of pulse experiments becomes more complicated, although transit times can still be obtained – a *knee* may be discerned in the current transient as the carrier front reaches the opposite electrode.

In ordinary polymers like polyethylene and poly(ethylene terephthalate) it is very difficult to observe any photoconduction at all, even when the sample is irradiated with UV radiation, where the photon energy is in excess of the large energy gaps of these polymers. This is a consequence of the very short life-time (typically 10^{-9} s or less) of carriers which suffer rapid recombination or deep trapping. A great experimental effort has gone into investigating the small degree of photoconduction that can be observed, but there is still uncertainty about the meaning of the results. When the pulse method is used, most workers report a *prompt* photo-current lasting less than 1 ns, followed by a much smaller current which decays as $1/t$. Some workers (e.g. Martin and Hirsch, 1972) argue that the prompt current corresponds to the transit of highly mobile (band-type) photogenerated carriers along polymer chains, the current ceasing as soon as the carriers reach chain ends. In several cases the prompt current has been definitely related to the injection of electrons from the metallic cathode by the external photoelectric effect (photon energy greater than the work function of the metal). This photoinjected current is strongly attenuated as the electrons are trapped by the polymer.

The relatively slow movement of the large concentration of trapped carriers near the photoinjecting electrode by a hopping process under the influence of the applied field would be expected to be space-charge limited. In some cases the latter part of the current transient does, indeed, show a break point indicating that at least some carriers eventually do make a complete transit to the opposite electrode. From this and other features of photoconduction transients, values of carrier drift mobilities ranging from $10^{-8}\,\mathrm{m^2\,V^{-1}\,s^{-1}}$ for poly(vinyl chloride) (Ranicar and Fleming, 1972) to $10^{-15}\,\mathrm{m^2\,V^{-1}\,s^{-1}}$ for polyethylene (Binks, Campbell and Sharples, 1970) have been obtained. These low values clearly confirm that some kind of hopping mechanism is responsible for the charge transport.

One class of organic polymer stands out above all others in its photoconductive efficiency. It is based on vinyl derivatives of certain polynuclear aromatic compounds, the most notable member being poly(*N*-vinyl carbazole),

commonly referred to as PVK. In pure PVK, conduction is dominated by holes which are long lived and readily photoinjected from metal electrodes. The unusual performance of this photoconductive polymer is undoubtedly connected with its tendency to take up a helical conformation with successive aromatic side chains lying parallel to each other in a stack along which electron transfer is relatively easy. Similar polymers, e.g. poly(2-vinyl carbazole) and poly(vinyl pyrene), are also good photoconductors, although the latter has poorer mechanical properties than PVK.

The photogeneration efficiency of PVK is greatly enhanced by the addition of an equimolar amount of the electron acceptor trinitrofluorenone (TNF) (Hoegl, 1965):

The resulting system of charge-transfer complexes strongly absorbs in the visible region of the spectrum to yield charge carriers. The conduction is then dominated by electrons rather than holes.

The actual mechanism of charge transport in PVK and PVK–TNP systems is not fully understood. The carrier mobilities are low, in the region of 10^{-10} m^2 V^{-1} s^{-1} for an applied field of about 1 kV m^{-1}, but they are strongly field and temperature dependent (Hughes, 1973). A field-assisted hopping process, analogous to field-assisted release of carriers from ionic centres in band conductors (Frenkel, 1938), is often suggested to explain the results.

PVK photoconductive systems have become important in the electro-reprographic industry. In xerography, for instance, the surface of a drum, coated with a photoconductive material, is first uniformly charged by a corona and then exposed to a bright image of the item to be copied. Where the light falls, the charge leaks away to the underlying earthed, metal drum. The remaining charged, originally dark, areas are able to attract the dry ink or *toner* for transfer to the copy paper. In the past amorphous selenium has been generally used as the photoconductive material on the drum, but PVK is now extensively used too. It has advantages in ease of fabrication and in toughness.

5.3.9   Superconduction

The subject of electrical conduction in organic compounds cannot be complete without mention of superconduction. Superconduction was first observed in metals by Kamerlingh Onnes in Leiden in 1911, soon after his success in liquefying helium. The phenomenon occurs when certain metals are cooled below a characteristic transition temperature where they spontaneously enter what is known as the superconducting state. In this state electrons can flow without any resistance at all. The technological potential of this is immediately apparent – lossless power transmission and powerful electromagnets would be obvious applications. The main snag is that the critical temperature for superconduction is always below 20 °K. When Little (1964) presented his ideas on a superconducting polymer, one of his most important conclusions was that organic polymers of specified structure might show superconduction up to very high temperatures, well above room temperature. In order to understand Little's prediction we ought to recapitulate on the explanation of conventional superconductivity.

The central feature of the generally accepted model of superconduction is the occurrence of Cooper electron pairs (Bardeen, Cooper and Schrieffer, 1957). Normally, electrons repel each other because like charges repel, but an effective attraction can develop between conduction electrons in certain metals at low temperature through distortion or polarisation of the atomic lattice (fig. 5.13). An electron polarises the

Fig. 5.13. Diagram of lattice polarisation caused by a passing conduction electron.

lattice around itself by its attraction for the metal atom cores which carry net positive charges. As the electron moves on, the lattice polarisation dies away but slowly; this trailing polarisation is attractive to another electron and at low temperatures is sufficient to provide an effective coupling between electrons. The situation is analogous to a pair of marbles rolling around on the top of a drum: they tend to follow each other because the second one tends to fall continually into the indentation caused by the first. The strength of the electron coupling in superconductors is inversely proportional to the square root of the mass of the positive ions of the lattice, so that if the ions are heavy, the displacement caused by a passing electron is small and vice versa.

Little envisaged a superconducting polymer with a conjugated backbone, the latter providing electrons in a 1-dimensional *running track*, and with dye-type side groups attached regularly along the chain (p. 123). Dye groups are characteristically electronically polarisable and Little suggested that formation of Cooper pairs amongst the π-electrons of the chain might occur through such polarisation of the side groups. Taking into account that the electronic mass is much smaller than any atomic mass, the coupling might be expected to be very large indeed in comparison with that in conventional superconductors. As a consequence of this, Little went so far as to predict a critical temperature for the transition from the superconducting to the normal state of about 2000 °K.

One reservation that one might have about actually observing superconduction in a polymer of this kind is that the gaps between molecules may preclude any bulk effect. However, in the metallic superconductors it has been found that supercurrents do flow across small gaps in an electrical circuit. The effect is called Josephson tunnelling (Josephson, 1962). It may therefore be that Josephson tunnelling would take care of the intermolecular problem.

Little's dramatic prediction of a room temperature, organic superconducting polymer has met with considerable criticism, and has not yet been substantiated in practice.

5.4   Conducting composites

Probably the simplest conducting composite consists of a fine metal powder dispersed uniformly throughout an insulating plastic matrix. The main drawback of this system, apart from the obvious *dilution* by the plastic of the conductivity of the metal, is that the metal particles remain isolated from each other, and therefore contribute no through-going conductivity to the composite at all, unless they are present in very high concentration. But high concentrations tend to destroy desirable mechanical properties of the plastic, the material typically becoming stiff and brittle. By way of illustration we can cite the results shown in table 5.1 for a dispersion of approximately spherical nickel particles (about 10 μm in diameter) in a low-density polyethylene. Here conductivity is lost entirely at concentrations below about 20% by volume of metal, corresponding to about 70% by weight. Nevertheless, conductive paints, which are frequently used for painting electrodes on to electrical test specimens and devices, work in just this way.

Table 5.1. *Conductivities of nickel–polyethy-lene composites*

Nickel content		Conductivity
% by weight	% by volume	$\Omega^{-1}\,m^{-1}$
70	19	10^{-6}
80	29	1.2×10
90	48	5.8×10^3

Composites based on silver powder can be made with conductivities as high as $10^6\,\Omega^{-1}\,m^{-1}$ at a loading of 85% by weight, when the insulating matrix, which may be an epoxy resin, serves essentially as a glue to hold the metal powder in position without altogether disrupting the metal–metal particle contact.

The *all or nothing* feature of metal powder composites is unsatisfactory for many applications, both because the poor mechanical properties incurred at concentrations where conduction is obtained cannot be tolerated, and because the high level of conductivity is not required or even not desirable. Thus a common requirement is for an antistatic material which has good plastics properties, sufficient conductivity to allow charges to leak away, and sufficient resistivity to prevent dangerous shocks to personnel who may become accidentally connected to mains electrical supplies through it.

The art of making a good conducting composite is to be able to use the minimum quantity of conductive component to achieve the required degree of electrical performance, and in this context it is important to know more about the factors which control the formation of conductive networks for a given concentration of conductive component. There are two main factors:

(*a*) *Quality of interparticle contacts.* Conductive networks in a 2-phase system depend on the particles of the conductive phase being able to make good electrical contact when they touch or come close to each other. Since a colossal number of interparticle contacts are involved, any changes in contact properties will be highly significant to the conductivity that may be realised. Oxidation of metal surfaces is detrimental to electrical contact and for this reason many metal powders (including copper) will not conduct even on their own unless highly compressed. The noble metals have the advantage in this respect, and some highly conducting composites employ silver-coated copper powder which combines the excellent contact properties of silver with the cheapness of copper.

(*b*) *Shape and size of conductive particles.* The resistivity of a compaction of carbon powder has been successfully explained on a simple model where the constrictions at interparticle contacts are assumed to have a dominating influence (Mrozowski, 1959). The theory uses Holm's (1946) expression $\rho\pi/a$ for the contact resistance between two spheres when the radius a of the circular area of contact between them is small with respect to the radius of the spheres, ρ being the resistivity of the material of the spheres. Thus, taking into account the way in which compressive forces cause elastic and yielding deformation, thereby altering a, the dependence of the resistivity of the compaction on pressure may be predicted quantitatively. The same theory may be expected to apply also to composites containing high loadings of conductive fillers where contacts amongst a certain proportion of particles form the conducting paths through the material. On this basis, where contact resistances dominate, spherical particles will be rather inefficient at imparting conductivity, since most of the material in the spheres is wasted as far as electrical conduction is concerned (see fig. 5.14). Relatively more is wasted when the sphere diameter is increased. Elongated shapes, especially fibres and even flakes, are much more efficient in this respect. The greater surface-to-volume ratio of these shapes also makes interparticle contact somewhat more likely at lower concentrations.

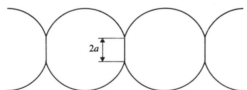

Fig. 5.14. Model form of intersphere contacts.

5.4.1   Conducting fibre composites

From the foregoing general considerations, it follows that a conductive additive will be efficiently deployed if it is distributed in such a way that long, interconnected chains exist permanently throughout the matrix. One obvious approach to this ideal is to start with a conductor which is already in fibrillar form, and fine metal wire or carbon fibres can be used in this way. Another variant is metal-coated fibre, silver on glass for example, where the metal is present as thin-walled tubes.

The electrical performance of a conducting fibre composite is somewhat analogous to the gelation of a polymer by crosslinking, the conducting network corresponding to the gel fraction. For a segment of fibre to be able to take part in electrical conduction it must be connected to the *gel* at both ends. At low fibre concentration there is no conducting network at all, and the conductivity of the composite is essentially that of the matrix material. As the fibre concentration is increased, the conductivity remains low until a 3-dimensional network is first established at the *gel-point,* when it suddenly increases by a large factor. Thereafter the conductivity changes more gradually with fibre concentration. The proportion of fibre which is conductively active continues to grow, however, since the number of interparticle contacts increases as the square of the fibre concentration.

The upper limit to the conductance between opposite faces of a composite block is the conductance of a uniform wire of the conductor, whose length equals the distance between the faces and whose volume equals the total volume of conductor in the block. In accord with this equivalent-wire limit, the maximum conductivity that can be obtained with a volume fraction f_v of a conductor (conductivity σ_c) in a composite is

$$\sigma_{max} = \sigma_c f_v. \tag{5.23}$$

In a composite containing randomly distributed fibres, orientation of fibre out of the field direction must effectively reduce the conductivity from the maximum. Consider a straight segment of fibre inclined at an angle θ to the applied field direction. Comparing the conductance of this segment with that of the same amount of conductor arranged uniformly, parallel to the field, its contribution to the cross-section of the equivalent wire is reduced by a factor $\cos^2 \theta$. For a random array of fibres the probability of an orientation in the range θ to $(\theta + d\theta)$ is proportional to $\sin \theta$, so that averaging over all orientations gives a mean reduction factor g thus:

$$g = \frac{\int_0^{\pi/2} \cos^2 \theta \sin \theta \, d\theta}{\pi/2} = \frac{2}{3\pi}. \tag{5.24}$$

If we assume that the fibre content is high, interfibre contacts will be sufficiently numerous for nearly all of the fibres to be active. Then, neglecting the effects of branch points and dead-ends, the equation for the conductivity of a composite containing randomly oriented fibres becomes

$$\sigma = \frac{2}{3\pi} \sigma_c f_v. \tag{5.25}$$

To cite a practical example, 20% by weight of 12 μm diameter glass fibre coated with silver (70 nm thick) mixed in a thermosetting polyester resin gave a composite conductivity of 2.5×10^4 Ω^{-1} m^{-1}. From equation (5.25) the calculated value is 5.3×10^4 Ω^{-1} m^{-1}, which compares reasonably well with the experiment, if one takes into account that some disintegration of the fibre is inevitable in the mixing process.

Preferential orientation of the fibres in a particular direction will tend, of course, to increase the conductivity in that direction at the expense of the others.

5.4.2   Carbon-black composites

The additive used more than any other to make conducting plastics is carbon black, which is produced by incomplete combustion of hydrocarbon vapours. It should be mentioned that the addition of particulate fillers has been used for a long time in the rubber industry primarily to give mechanical reinforcement, and carbon black has generally been preferred for this, because (*a*) it is a compatible material, mixing in, and adhering to, the matrix rather well, (*b*) it does not change the overall density very much, and (*c*) it is cheap. Rather higher concentrations of special grades are necessary, however, for electrical purposes. The dependence of conductivity on concentration of a typical carbon black is shown in fig. 5.15. The upper limit of conductivity that can be achieved is about 100 Ω^{-1} m^{-1}.

Particles of carbon black are very small, with diameters in the range 10 to 300 nm, and densities of about 1.8 Mg m^{-3}. They consist of roughly spherical clusters of small fragments, each of which is similar to a small bit of graphite (a few parallel layers), but with the hexagons in successive layers oriented in different directions. Although the shape of the graph of log (conductivity) against carbon-black concentration is similar for all grades of carbon black, the conductivity for a given concentration depends very much on the way the black has been made. The so-called *furnace* blacks give the highest values. Blacks which give high conductivity are characterised by small particle size, but, more importantly, by a propensity to agglomerate into chains to give a pseudo-fibre arrangement, usually referred to as *structure,* which may persist even when the black is mixed into another material. The reason for this structure is not fully understood. Some of it may be explained by the amalgamation of droplets of high-molecular-weight hydrocarbons into *necklaces* at an intermediate stage in the manufacture, but this does not explain the fact that some structure which is disrupted by shearing action will reform on resting. The condition of the surface of a carbon black affects conductivity, probably through its influence on structure. Thus treatment of

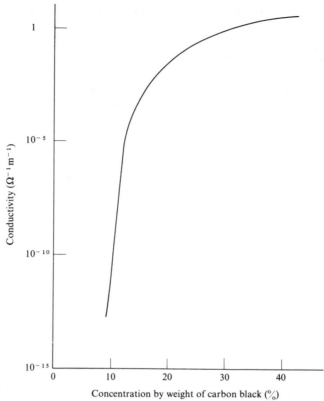

Fig. 5.15. The dependence of the conductivity of a poly(vinyl chloride) composite on carbon-black content.

particles to remove surface hydrogen and oxygen greatly increases the conductivities that may be obtained in composites. An indication of the degree of structure inherent in a particular black can be obtained by various empirical methods. The most common index of structure is oil absorption, defined as 'the minimum volume of oil (usually dibutyl phthalate) which will give, under conditions of controlled mixing, a mix having no voids'. Spherically shaped powders pack closely whereas branched-chain aggregates in a high-structure black give rise to more voids and thereby to a high absorption of oil.

In the face of the complicated nature of carbon black it is not surprising that any theoretical relationship between the resistivity of a composite and its carbon-black content is difficult to obtain. For one thing, the all-important structure is affected by the process of mixing the black into the polymer. Nevertheless there are two theories which go some way in explaining the behaviour of the composites in certain regimes.

The first theory (Voet, Whitten and Cook, 1965) sets out to treat those systems which show non-Ohmic conduction. This kind of behaviour is indeed often found at the low-conductivity end of the scale, and it implies that some kind of electron emission process controls the conduction, probably a tunnelling of electrons from one particle to the next across gaps of up to 5 nm, say. It is relevant to note here that the very small particle size of carbon blacks means that Brownian motion might be responsible for momentarily bringing particles close enough together for electron tunnelling to occur. Emission current density j is related to applied voltage V by an equation of the form

$$j = AV^n \exp(-B/V), \tag{5.26}$$

where A, B and n are constants; n usually lies between 1 and 3. A is a function of the tunnelling frequency, i.e. the *attempt-to-escape* frequency of the electron, and the factor $\exp(-B/V)$ defines the transition probability through the energy barrier of the gap. B will be proportional to the gap between particles, so that, neglecting the pre-exponential factor, the effective conductivity σ at a particular voltage will be given by

$$\ln\sigma \propto 1/\bar{l}, \tag{5.27}$$

where $\bar{l}$ is some average interparticle separation. Now it may be shown that the interparticle separation for an array of spherical particles (radius a) on a cubic lattice (repeat distance l) where $l \gg a$ is given by

$$l = a\left(\frac{4\pi}{3f_v}\right)^{\frac{1}{3}}, \tag{5.28}$$

where f_v is the volume-fraction of particles. For a random, low-density dispersion of particles we may therefore expect that

$$\bar{l} \propto f_v^{-\frac{1}{3}}. \tag{5.29}$$

Combining this with equation (5.27) we then obtain

$$\ln\sigma \propto f_v^{\frac{1}{3}}. \tag{5.30}$$

To assess the validity of this equation we can compare it with the purely empirical equation (Bulgin, 1945) which has a similar form:

$$\ln\sigma = \left(\frac{A}{f_w}\right)^{-p}, \tag{5.31}$$

where f_w is the weight-fraction of the carbon black in the composite (for low carbon-black concentrations $f_w \propto f_v$), and A and p are constants.

Most experimental data may be fitted by the latter equation, although the constants A and p vary for different kinds of blacks. Values of p in fact lie in the range 0.6 to 5.0 (Norman, 1970), so that we can only say that the theoretical equation (5.30) gives rather poor quantitative agreement with experiment.

The second theory, due to Scarisbrick (1973), treats the high-conductivity cases where Ohmic behaviour is encountered. He tacitly assumes that interparticle contacts are Ohmic, and goes on to calculate the probability of formation of conductive chains. Supposing the composite to consist of a random mixture of conducting and non-conducting elements, the probability of obtaining a sequence of n conducting elements is f_v^n. If the average conductive element is a sphere, diameter d, associated with a spherical region, diameter D, of the composite:

$$\frac{D}{d} = \frac{1}{f_v^{\frac{1}{3}}}.\tag{5.32}$$

By random-walk theory, if the diameter D is taken as the mean distance between the ends of a chain of n links of length d,

$$D = dn^{\frac{1}{2}}.\tag{5.33}$$

Combining equations (5.32) and (5.33),

$$n = f_v^{-\frac{2}{3}}.\tag{5.34}$$

Hence the probability that the volume associated with one conductive element may be bridged by chain formation is then given by

$$p = f_v^{f_v^{-\frac{2}{3}}}.\tag{5.35}$$

Scarisbrick also draws attention to the geometrical factor which connects the observable conductivity with the random assembly of conducting chains. Fig. 5.16 shows conducting chains *resolved* along the axial directions of a unit cube, and collected together to give a cross-sectional area C^2 in each of the three directions. Taking into account the mutual volume, where the chains along each direction overlap,

$$f_v = 3C^2 - 2C^3.\tag{5.36}$$

The ratio of the conductivity of the composite to that of the conductive component itself is the product of the proportion of conducting to non-conducting elements, the probability of chain formation and the geometrical factor, i.e.

$$\frac{\sigma}{\sigma_c} = f_v \times f_v^{f_v^{-\frac{2}{3}}} \times C^2.\tag{5.37}$$

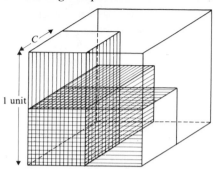

Fig. 5.16. Equivalent distribution of conductive chains along the principal directions of a unit cube of an isotropic conducting mixture, defining the geometrical factor *C* (Scarisbrick, 1973). Copyright of the Institute of Physics.

Comparison of this result with experimental values obtained for a dispersion of carbon black in polyethylene is shown in fig. 5.17, where the agreement is seen to be fairly good, the discrepancy being at most a factor of about three.

Neither theory takes the macrostructure of carbon black into account in any quantitative way, although in practice this is crucial in selecting a material for a particular application.

One way which is often used to impart an artificial structure to the dispersion of a carbon black in a composite is to coat the particles of a moulding powder, e.g. 1 mm diameter particles of polyethylene, with a conductive carbon black by mixing them together in a ball-mill. If subsequent moulding does not shear the mixture too much (rotational casting is ideal) the black remains concentrated in a honeycomb-like network, and the overall conductivity is greatly enhanced over that of a uniform dispersion of the same proportion of black.

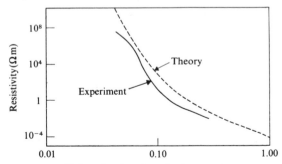

Fig. 5.17. Comparison of the experimental and theoretical dependences of resistivity on carbon-black content in a polyethylene composite (Scarisbrick, 1973). Copyright of the Institute of Physics.

Some carbon-black composites are used as heating elements. In this application the operating temperature may be as high as 150 °C, so that the temperature dependence of the conductivity of the product is of the utmost importance for stable working. If the temperature coefficient of conductance were positive there would be the danger of runaway conditions developing. Although the temperature coefficient of conductance for graphite is in fact positive, the temperature coefficient of a composite may be positive or negative depending primarily on the degree to which conductive chains are broken up by thermal expansions and contractions. The outcome in the case of a silicone-rubber composite, a suitable one for heating panels, is a coefficient of -0.0003 deg^{-1}.

Any mechanical strain can upset the conductive-chain formation and pronounced effects of this kind are observed in conductive rubbers (Norman, 1970).

In summary, carbon blacks work well as conductive additives for plastics and rubbers for applications other than power transmission, provided that a high-structure black and a suitable set of processing conditions are chosen, and as long as the black colour is acceptable.

5.5   Measurement of resistivity

In this section we shall consistently refer to resistivity, rather than to its reciprocal, conductivity. The choice is arbitrary and commercial instruments are sometimes calibrated in units of resistance and sometimes in units of conductance.* Unless otherwise stated, when we refer to resistivity we shall mean volume resistivity ρ_v (Ω m), the resistance between opposite faces of a unit cube. A surface resistivity ρ_s is often used to characterise current flow over a surface, as with an antistatic coating, and is defined as the resistance between opposite edges of a unit square. We note first that the resistance across a square is independent of the size of the square and that the unit of surface resistivity is simply the ohm (Ω), occasionally written rather superfluously as ohm per square. Secondly, a conducting surface must in reality be a layer with a finite thickness t, and we have only an *effective* surface resistivity, which is related to the true volume resistivity of the layer by

$$\rho_s = \frac{\rho_v}{t}. \tag{5.38}$$

In a similar way effective surface resistivities of homogeneous sheets are often quoted.

There are two simple types of specimen and electrode arrangement

* The unit of conductance, the reciprocal ohm (Ω^{-1}), is often called the siemen (S).

that are basic to volume-resistivity measurement. One is a rectangular or cylindrical block with electrodes on the ends. The other is like that used for measurements of dielectric constant where electrodes are applied to either side of a thin disc. The latter is more appropriate to high-resistivity materials. In both cases the volume resistivity is related to the measured resistance R between the electrodes by

$$\rho_v = \frac{RA}{l},$$ (5.39)

where A and l are the cross-sectional area and the length of the specimen between the electrodes, respectively.

Concentric ring electrodes are the easiest to use for measurement of surface resistivity. The resistance R between them is the sum of the resistances of the elemental annuli in series:

$$R = \int_{r_1}^{r_2} \frac{\rho_s}{2\pi r}\, dr.$$ (5.40)

where r_1, r_2 are the radii of the inner and outer electrodes, respectively. Hence

$$\rho_s = 2\pi R \ln\left(\frac{r_2}{r_1}\right).$$ (5.41)

The main problem in accurate measurement of a low resistivity is one of contact resistance between the measurement electrodes and the specimen. Contact resistance may be reduced by painting electrodes directly on to the surface of the specimen instead of relying on pressure contact with metal plates or foils. Suitable paints are silver dispersions or Aquadag (an aqueous dispersion of colloidal graphite). A much better way to deal with contact resistance is to use a 4-terminal, potentiometric method, as illustrated in fig. 5.18(a). A known current density j is established in the central region by a battery connected to the outer electrodes. If the electric field E is then determined by measuring the potential drop ΔV across two inner electrodes, separation Δx, the resistivity is given by

$$\rho_v = \frac{j}{E} = \frac{I/A}{\Delta V/\Delta x}.$$ (5.42)

Any effect of contact resistance is then avoided, providing the contact resistance is much smaller than the input resistance of the voltmeter.

When making measurements of a high resistivity the main problem becomes one of leakage of current between the input terminals of the

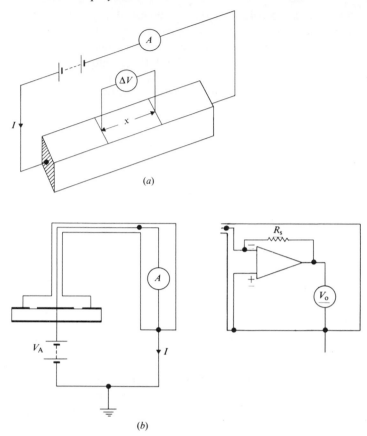

Fig. 5.18. Circuit diagrams for resistivity measurements: (*a*) 4-terminal method, (*b*) 2-terminal method with guarding.

ohmmeter via routes other than the intended one through the specimen. Surfaces often provide a low-resistance path through the accumulation of dirt and moisture on them. One can usually overcome the problem to a large extent by the use of an extra guard electrode on the specimen. Any leakage currents over the surface of the specimen are then collected by the guard electrode and are not included in the measured current. In its most convenient form (see fig. 5.18(*b*)) the guard is connected to earth and surrounds that electrode which is connected to the highly insulated side of the ohmmeter, whilst the other electrode is maintained at a fixed, high voltage V_A. The guard may be extended to shield the connecting lead too. So long as the guarded electrode itself stays near earth potential, there will be no significant current flow between it and the guard; this will appertain if the internal resistance of the ohmmeter is much less than the sample resistance. In extreme cases, where the sample resistance is very high, the current I which flows is very small and it can only be determined by measuring the voltage drop across a standard high resistor R_s with an electrometer (see fig. 5.18(*b*) again). In that case the effective input resistance of the electrometer may be kept relatively low, so as to maintain the virtual earth condition at the input, by connecting the standard resistor in a negative-feedback loop of the high-gain $(-N)$, high-input-resistance amplifier. If the output of the amplifier, measured by a relatively low-resistance voltmeter, is V_o,

$$I = V_0/R_s, \tag{5.43}$$

and

$$R = R_s\, V_A/V_0. \tag{5.44}$$

In the foregoing methods of resistivity measurement specially shaped specimens and electrodes are required, the arrangements being chosen to give uniform, or uniformly divergent, fields in the material. An alternative approach which is useful for rapid measurements on small specimens or for *in situ*, non-destructive tests on a moulded item is to use some kind of electrode probe applied to the outside of the material in question. When pointed electrodes are used in this way, however, the electric fields and current densities are no longer uniform. Fortunately, there is a strict analogy between the field equations of current flow and those of electrostatic charges, since they both comply with Laplace's equation:

$$\frac{\partial^2 V}{\partial x^2} + \frac{\partial^2 V}{\partial y^2} + \frac{\partial^2 V}{\partial z^2} = 0, \tag{5.45}$$

where V is the potential at any point in the field. Thus electrodes which inject or drain current are equivalent to positive and negative charges. For example, the potential at a distance r from a point electrode supplying a current I inside a material is given by

$$V = \frac{I\rho_v}{4\pi r}, \tag{5.46}$$

which may be compared with the potential due to a point charge q in a dielectric medium:

$$V = \frac{q}{4\pi\varepsilon_0\varepsilon' r}. \tag{5.47}$$

We may therefore make use of well-known solutions of electrostatic equations in order to relate resistance measured by a point probe to volume resistivities.

Suppose that a pair of hemispherical electrodes makes contact with the surface of a semi-infinite sample in the way shown in fig. 5.19(a). If the separation of the electrodes d is much greater than the hemispherical radius r_0, analogy with the field equations for charged spheres (Moon and Spencer, 1961) shows that the measured resistance R between the electrodes (neglecting contact resistances) will be given by

$$R = \frac{\rho_v}{\pi r_0}(r_0 \ll d). \tag{5.48}$$

Here we see that the resistance is independent of the electrode separation. The physical reason is that the major part of the voltage drop occurs in the immediate vicinities of the tips of the electrodes, and we

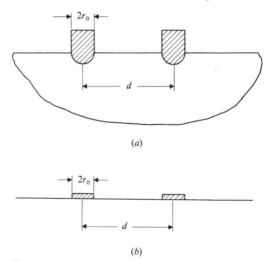

(a)

(b)

Fig. 5.19. Types of 2-terminal resistance probes: (a) hemispheres in a conductive solid, (b) discs on a conductive surface.

may think of the resistance as the sum of a *spreading* resistance from the source electrode and a *convergence* resistance to the drain electrode. From this we conclude that 2-point probe measurements will not be very reliable or reproducible, because they will depend so sensitively on the small sample of material at the electrode tips. On a pliable material the form of indentation and contact area will be especially difficult to control with precision, and, in addition, we recognise that the material will be most distorted by the electrode pressure in just that region to which the resistance measurement is most sensitive. Therefore 2-point volume-resistivity measurements are not recommended.

The resistance R measured on a conductive surface by a 2-point probe will depend both on the radius r_0 of the circular contact at each electrode and on their separation d (see fig. 5.19(b)):

$$R = \frac{\rho_s}{\pi} \cosh^{-1} \frac{d}{2r_0}. \tag{5.49}$$

Unlike the case of volume resistivity, the resistance never becomes independent of electrode separation. Again measurement with finely pointed electrodes will be over-sensitive to the precision of contact.

The situation is quite different with 4-point probe measurements, shown in fig. 5.20. In this type of measurement, a known current I is

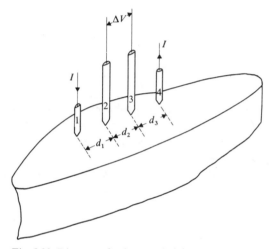

Fig. 5.20. Diagram of volume resistivity measurement with a 4-point probe.

injected by electrode 1 and collected at electrode 4, whilst the potential difference ΔV between electrodes 2 and 3 is measured. Using equation (5.46) for the potential of a current source, one can obtain the net potentials at electrodes 2 and 3 due to the current electrodes on the semi-infinite block, whence (Valdes, 1954)

$$\Delta V = \frac{I\rho_v}{2\pi} \left(\frac{1}{d_1} - \frac{1}{d_2 + d_3} - \frac{1}{d_1 + d_2} + \frac{1}{d_3} \right). \tag{5.50}$$

If the electrodes are equispaced distance d apart, equation (5.50) reduces to

$$\Delta V = \frac{I\rho_v}{2\pi d}. \tag{5.51}$$

The result is independent of the electrode contact area if the size of the contacts is much smaller than the inter-electrode spacing.

When the electrodes are applied to a specimen with finite boundaries, the apparent resistivity ρ_v' which is measured is higher than the true value. This is usually expressed in terms of a correction divisor (CD):

$$\rho_v = \frac{\rho_v'}{CD}. \tag{5.52}$$

Uhlir (1955) has derived expressions for CD for various geometries by the method of images. For a conductive sheet of thickness t where $t \ll d$ the expression for CD reduces to

$$CD = 2\frac{d}{t}\ln 2. \tag{5.53}$$

Equation (5.51) for the 4-point probe now becomes

$$\Delta V = \frac{I\rho_s \ln 2}{\pi d}. \tag{5.54}$$

A more general treatment of 4-point probe measurements which includes alternative electrode arrangements and takes into account anisotropy of the material, has been given by van der Pauw (1961).

4-point probe measurements are very reliable and they have been used extensively for measurements on inorganic semiconductors. The technique has limitations, however, for as more highly resistive materials are examined the point contacts are incapable of supplying currents that are high enough to make ΔV readily measurable. Furthermore, the necessary input resistance of the voltmeter, which must be greater than the resistance between points 2 and 3, becomes very high. The practical upper limit of resistivities that can be measured by the 4-point probe technique is about $10^8 \, \Omega$ m.

Observation of the rate of decay of charges by leakage to earth has formed the basis for other methods of determining high resistivities, especially surface resistivities of film or sheet material. The specimen is first charged to a high voltage by one of several methods: (a) exposing the surface to a source of air ions, (b) rubbing with another material to produce charge transfer, or (c) connecting the edges to a high voltage source. The charge is then monitored as a function of time by means of an electric-field meter (see § 7.2). The decay rate is difficult to interpret exactly, for it depends on the product of the system's effective resistance and capacitance to earth; these are related in a rather complex way to the geometrical arrangement employed (Henry, Livesey and Wood, 1967). As a rough guide, a time constant (assuming an exponential decay) of one second corresponds to a surface resistivity of $10^{11}\ \Omega$ in most practical situations.

5.6   Further reading

The book on *Organic Semiconductors* by Gutmann and Lyons (1967) develops the basic theory of conduction processes relevant to polymers and presents a large collection of data. Specific aspects are treated in detail in the books edited by Katon (1968) and by Frisch and Patsis (1972). Conducting composites have been covered by Norman (1970).

6 Dielectric breakdown

6.1 Introduction

If the voltage across a piece of dielectric material is steadily increased, there must come a point when any imperfections in the insulating properties of the material or its surroundings will become apparent, and total breakdown will eventually ensue. Characteristically, the final event is localised, sudden and catastrophic. At the high voltage involved the quick release of so much electrical energy usually means that the material burns out in the breakdown region between the electrodes. Although dielectric breakdown, like mechanical failure, is invariably connected with localised imperfections or weaknesses of some kind, we still try to define a relevant material property. Thus the existence of a maximum voltage which an insulator will support for a long time without failing leads to the concept of a dielectric strength, defined as the breakdown voltage divided by the thickness of the insulator, i.e. a maximum electric *field* which the *material* can sustain indefinitely. The intrinsic dielectric strength of a homogeneous solid is evidently very high, usually in excess of 100 MV m^{-1}, and proves to be a very elusive fundamental property. The reason for this is that a particular specimen may often more easily fail in many different ways which have more to do with its environment, its physical state and purity and the type of electrode used, than with its basic constitution. These alternative breakdown mechanisms are the ones which generally limit the effective strength of an insulator in a practical situation and they are difficult to avoid altogether. The quest for an intrinsic strength has consequently become somewhat academic. In this context we must beware standard tests of *dielectric strength* of solids – they usually do not in fact measure an intrinsic value, because they allow premature discharges to occur in the surrounding gaseous or liquid medium. The main approach that has been adopted in industry for the necessary assessment of breakdown behaviour of materials in exercises of product improvement and replacement has been to design special tests. These simulate the conditions of the applications concerned. In this context the ageing or deterioration of the material during service, e.g. chemical degradation in strong sunlight and cracking under prolonged mechanical stress, is vitally important too, since changes brought about in this way almost always introduce electrical weaknesses.

In the following sections we discuss briefly the principal mechanisms of electrical breakdown in solid polymers and some of the consequences for the use of polymers as dielectric materials.

6.2   Electronic breakdown

Insofar as there is such a thing as an intrinsic breakdown strength of an insulating material, the most likely source of the instability will be the small numbers of electrons which are available for acceleration by the applied field. By analogy with the mechanism of sparking in gases, one can imagine that a Townsend-like avalanche may occur whenever the field is high enough for a conduction electron to gain sufficient energy to excite more electrons by collision. Actually, the breakdown strengths calculated for insulating solids on the basis of the simple gas model for the cumulative process are much too high. More successful approaches take into account the high electron mobilities in regular lattice structures, and are rooted in band theory.

In one of the first theories of electronic breakdown von Hippel considered the situation in a pure, crystalline material at low temperatures. In such a system electron–electron collisions are rare, and electrons are principally scattered by interaction with the vibrations of the lattice (electron–phonon collisions). Continual thermalisation of the electrons by this process sets the limit to the kinetic energy that an electron can acquire through acceleration in the electric field. One immediate prediction is that a rise in temperature, which enhances the scattering of electrons by the lattice (the mean free path of electrons between collisions with phonons is shortened) will increase the rate at which electrons lose the energy that they have gained from the electric field. The dielectric strength will therefore *increase* with temperature. Careful measurements at low temperatures on simple ionic compounds, e.g. sodium chloride, have largely substantiated this.

At high temperatures and in crystals containing many impurities and imperfections, as well as in molecular and amorphous solids, including polymers, Fröhlich recognised that the combined number of electrons in localised excited states and in a conduction band, if it exists, will be relatively high, so that electron–electron scattering must predominate. Since the transfer of energy from these electrons to the lattice is rather slow, the electron temperature T may rise above the lattice temperature T_0 when an electric field feeds energy directly to the conduction electrons. Since we only have rather limited knowledge of conduction processes in polymers, a good model for detailed calculations is not available and we can only make a reasonable guess at a generally

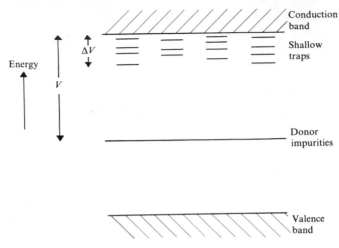

Fig. 6.1. Model energy-level scheme for a polymer.

appropriate one. Fröhlich (1947) based his calculations on the hypothesis of the energy-level scheme shown in fig. 6.1, where conduction electrons are derived from impurity levels lying deep ($V = 1$ eV or more) in the forbidden zone. There is also a set of shallow traps spread below the conduction-band edge ($V \gg \Delta V \gg kT$). In outline, the theory of breakdown is then as follows.

In an applied electric field E, energy is transferred directly to the conduction electrons (charge e, mass m) at a rate $A = jE$, where j is the current density. If we suppose that each electron is accelerated for a time 2τ, on average, between collisions at which its energy is completely randomised, then the mean drift velocity of the conduction electrons in the field direction is $-(eE/m)\,\tau$. The current density due to drift of all the conduction electrons (concentration n) is $ne^2E\tau/m$, and we consequently have

$$A = \frac{e^2\tau n E^2}{m}. \tag{6.1}$$

From semiconductor theory the concentration of conduction electrons is given by

$$n = C(T)\mathrm{e}^{-V/2kT}, \tag{6.2}$$

where $C(T)$ is a factor which depends on the effective density of states in the conduction band and which varies only slowly with temperature. Substituting for n in equation (6.1),

$$A = \frac{e^2\tau E^2}{m}C(T)\mathrm{e}^{-V/2kT}. \tag{6.3}$$

The energy thus gained from the field will be rapidly shared amongst the conduction and trapped electrons by electron–electron collisions.

Let us now consider the loss of energy from the electrons to the molecular lattice. The concentration of trapped electrons will be given approximately by

$$n_T = C_T(T)e^{-(V-\Delta V)/2kT}, \tag{6.4}$$

where the factor $C_T(T)$ now relates to the density of trapping states and again varies slowly with temperature. If we assume a high density of traps, the concentration of trapped electrons will greatly exceed the concentration of conduction electrons, and it will be mainly trapped electrons which are responsible for transferring energy to the lattice.

For simplicity, suppose that energy is transferred to a single vibrational mode of frequency v in single phonon jumps of energy hv. The number of transitions per unit volume per second W_e of an electron from a level of energy U to one of $U+hv$ involving phonon emission by the lattice is the product of the electron transition probability $P(U)$, the concentration $f(U)$ of electrons in levels of energy U, and the average number density $N_v(T_0)$ of vibrational quanta (phonons) present in the lattice at temperature T_0, i.e.

$$W_e = P(U)f(U)N_v(T_0). \tag{6.5}$$

Similarly, for transitions in the reverse direction with phonon absorption:

$$W_a = P(U)f(U+hv)[N_v(T_0)+1]. \tag{6.6}$$

Hence, the rate of energy loss B per unit volume to the lattice from trapped electrons is

$$\begin{aligned}B &= hv\Sigma(W_a - W_e) \\ &= hv\Sigma P(U)\{f(U+hv)[N_v(T_0)+1]-f(U)N_v(T_0)\},\end{aligned} \tag{6.7}$$

where the summation extends over all trap levels. Assuming that an equilibrium exists in the trap population,

$$\frac{f(U+hv)}{f(U)} = e^{-hv/kT}. \tag{6.8}$$

Then, replacing $P(U)$ for all levels by a mean value P, noting that $\Sigma f(U) = n_T$, and using the Planck expression $(e^{hv/kT_0}-1)^{-1}$ for $N_v(T_0)$, we obtain from equation (6.7)

$$B = hvP(e^{hv/kT}-1)^{-1}C_T(T)e^{-(V-\Delta V)/2kT}(e^{hv/kT_0-hv/kT}-1). \tag{6.9}$$

Taking into account that τ and P will depend only weakly on tempera-

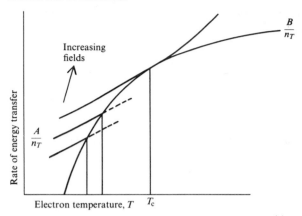

Fig. 6.2. Energy-transfer curves for increasing applied fields, showing the onset of electron temperature runaway. After Fröhlich (1947), by courtesy of the Royal Society.

ture, equations (6.3) and (6.9) show that the rates of energy gain and loss per trapped electron, given by A/n_T and B/n_T respectively, vary with temperature in the manner indicated in fig. 6.2, where a family of energy-transfer curves for different applied fields is shown. At low field strengths, an equilibrium electron temperature will be attained, energy lost to the lattice just balancing energy gained from the field; but at fields above a critical value E_c, no balance is possible and this defines the breakdown field strength. Substituting for the critical condition in the energy balance equation ($A = B$), and taking the first approximation in exponential expansions, we obtain the relation

$$E_c \propto e^{\Delta V/4kT_0}, \tag{6.10}$$

which predicts a decrease in breakdown field strength as temperature rises. The physical reason for such behaviour with this model is the increase in the relative number of conduction, as opposed to trapped, electrons which can gain kinetic energy from the field. Von Hippel and Maurer (1941) found confirmatory evidence of this effect in fused (amorphous) quartz, in sharp contrast to the results for pure crystalline quartz (fig. 6.3).

As we intimated in the introduction, it is arguable whether truly intrinsic breakdown has ever been observed in polymers. In this sense an intrinsic breakdown strength must represent an upper limit to any value that can ever be realised experimentally. The most reliable measurements, judged by the large values which were obtained in comparison with those from other measurements, have been made with recessed specimens, illustrated in fig. 6.4, with evaporated aluminium electrodes.

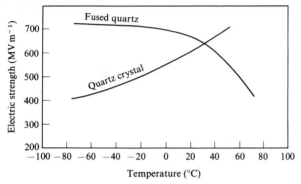

Fig. 6.3. Comparison of breakdown behaviour of crystalline and amorphous (fused) quartz (von Hippel and Maurer, 1941).

The recess design neatly places the high stress just where it is required across a thin layer, whilst at the same time avoiding excessive stresses in the air or other medium surrounding the edges. Results for polyethylene (Lawson 1966), shown in fig. 6.4, definitely indicate a fall in dielectric strength with temperature in accord with Fröhlich's theory. The extra, steep fall at higher temperatures is probably not electronic in origin as we shall see in the next section. Somewhat higher values still for the

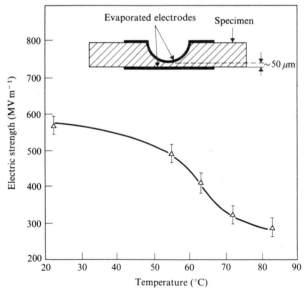

Fig. 6.4. The variation of the electric strength of polyethylene with temperature (Lawson, 1966). The inset diagram shows the type of recessed specimen used.

breakdown strength of low-loss polymers have been obtained with a more complicated arrangement, where the specimen is embedded in epoxy resin (McKeown, 1965), but there is some uncertainty as to whether this is an experimental artefact or not.

In order to predict absolute dielectric strengths we should need to have more detailed information than is yet available about electronic states and mobilities in polymers. For the present we can only conclude that there is satisfactory agreement between the form of the theoretical results, based on a rather general electronic model, and the best experimental results. To the extent that the model is a very reasonable one, we can say that we can understand intrinsic breakdown behaviour.

Measurement of pre-breakdown currents, especially with pointed electrodes which impose regions of very high field strength at their tips when embedded in the material, suggests that electronic carrier injection from the electrodes (Schottky emission) or from impurities (Poole–Frenkel effect) may play a part in the breakdown process in some cases. More work is required, however, before this can be fully understood.

6.3   Electromechanical breakdown

Electrodes attached to the surface of a specimen during a dielectric breakdown test will exert a compressive force on the specimen by mutual Coulombic attraction of the electrodes as the voltage V is imposed. If this is sufficient to cause appreciable deformation at fields below the intrinsic breakdown value, the dielectric strength will be reduced. The attractive force F is given by the differential of the energy U stored in the system with respect to the thickness d of the material at constant applied voltage:

$$F = \left(\frac{\partial U}{\partial d}\right)_V = \frac{\partial}{\partial d}\left(\tfrac{1}{2}CV^2\right)_V. \tag{6.11}$$

Substituting for the parallel-plate capacitance C of the specimen (cross-sectional area A, dielectric constant ε'), we obtain an expression for the compressive force per unit area:

$$\frac{F}{A} = -\tfrac{1}{2}\varepsilon_0\varepsilon'\left(\frac{V}{d}\right)^2. \tag{6.12}$$

At equilibrium the electrical force causing compression is balanced by the elastic restoring force, as expressed by the equation

$$\tfrac{1}{2}\varepsilon_0\varepsilon'\left(\frac{V}{d}\right)^2 = Y\ln\left(\frac{d_0}{d}\right), \tag{6.13}$$

where Y is Young's modulus for the material, and d_0 is the initial thickness of the specimen. Solving this equation for d then gives the equilibrium thickness of the slab under the applied voltage. Since $d^2\ln(d_0/d)$ has a maximum value when $d/d_0 = \exp(-1/2) \approx 0.6$, no real value of V can produce a stable situation for values of d/d_0 less than 0.6. Further increase of V produces mechanical collapse, the critical electric stress being

$$E_c = \left(\frac{Y}{\varepsilon_0\varepsilon'}\right)^{\frac{1}{2}}. \tag{6.14}$$

The highest apparent dielectric strength E_a which can be observed is therefore

$$E_a = \frac{V_c}{d_0} = \frac{d}{d_0}E_c \approx 0.6\left(\frac{Y}{\varepsilon_0\varepsilon'}\right)^{\frac{1}{2}}. \tag{6.15}$$

(We should note that the above simple treatment totally ignores any departure from linear elastic behaviour at large strains.)

Stark and Garton (1955) recognised this behaviour in polyethylene when they compared ordinary polyethylene with an irradiated sample whose Young's modulus (and also dielectric strength) did not decrease so much with temperature.

Apparently low values of breakdown strength of many rubbery materials are described quantitatively by equation (6.15), and most plastics fail by the electromechanical mechanism at high temperatures.

6.4   Thermal breakdown

Whenever there is sufficient conductivity present in a dielectric to produce appreciable Joule heating in an applied field, the possibility of thermal runaway exists, for the accompanying rise in temperature will increase the conductivity still further. In alternating fields there may be additional heat generated through one or more relaxational processes as described in chapter 3, and this will hasten the onset of any thermal runaway condition. Whether thermal breakdown will eventually develop in this way or not will also depend on the rate at which heat is conducted away to the surroundings. The heat balance equation is expressed by the following continuity equation:

| Electrical power dissipated in material per unit volume | $=$ | Rate of increase in heat content | $+$ | Rate at which heat is conducted away |

i.e.

$$\sigma E^2 = C_v \frac{dT}{dt} - \text{div}\,(\kappa\,\text{grad}\,T), \tag{6.16}$$

where σ is the electrical conductivity of the material, C_v is its heat capacity per unit volume and κ is its thermal conductivity. It is not possible to obtain a general solution of this equation, because σ, C_v and κ are all functions of temperature. Approximate numerical solutions have been obtained, however, for a semi-infinite dielectric slab between electrodes maintained at a constant temperature, and these predict that the temperature of the hottest part of the dielectric at the centre of the slab varies with time for different constant applied voltages in the way shown by the curves in fig. 6.5 (Whitehead, 1951). The main feature is a critical voltage V_c above which the temperature increases indefinitely. This will reduce the intrinsic dielectric strength until breakdown ensues, unless melting or chemical decomposition intervenes.

We can appreciate that the critical voltage V_c will be independent of thickness in this particular case, because an increase in thickness will decrease the electric field for a given applied voltage, but reduce the rate at which heat can escape from the central region by conduction. For this reason thermal breakdown *voltages* are sometimes quoted. Thermal breakdown in polymers is most liable to occur (*a*) at high ambient temperatures where electrical conductivity may become appreciable, e.g. Nylon-6,6 above 90 °C, or (*b*) at high frequencies where the rate of heat dissipation by dipolar relaxation processes may be high.

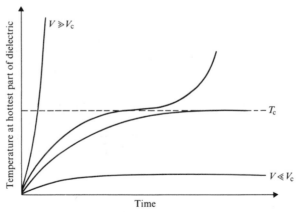

Fig. 6.5. Calculated heating curves for increasing voltages applied across an infinite slab, showing the onset of thermal runaway (Copple *et al.*, 1939).

6.5   Breakdown caused by gas discharges

The dielectric strength of a gas is very much less than that of a solid insulator, being of the order of 3 MV m^{-1}. Consequently, during the application of a high voltage to a solid specimen, discharges are likely to occur at an early stage in any gas which is at the edges of the electrodes or may be occluded as bubbles in the solid. Such external or internal discharges tend to damage the solid and repeated discharges lead to dielectric failure. Organic polymers are especially prone to this type of failure, because the high temperatures in the discharges readily cause degradation to carbon and gaseous products. (The process is often accelerated by reaction with ozone and other active species formed in the gaseous discharges.) With an applied direct voltage, discharges will recur only after surface charges deposited by the previous discharge have had time to leak away, but under an alternating voltage, discharge will be repeated with every half-cycle and the problem becomes much more severe.

Before we can recount the effects of gas discharges in more detail we need to say something about the nature of gas discharges themselves. In a uniform field the breakdown is a cumulative process and spark voltages follow Paschen's law, which says that the minimum voltage V_g necessary to produce a spark across a gap depends on the product $l\rho$ where l is the length of the gap and ρ is the density of the gas. The Paschen's law curve for air is shown in fig. 6.6. From the latter we may deduce that (*a*) at constant density the breakdown field strength increases towards shorter gaps, and very steeply for gaps smaller than 1 mm at atmospheric pressure and (*b*) for a given gas there is a minimum

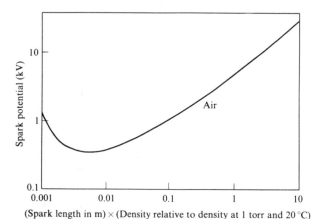

(Spark length in m) × (Density relative to density at 1 torr and 20 °C)

Fig. 6.6. Paschen's law curve for air from *Dielectric Breakdown of Solids* by S. Whitehead, published by Oxford University Press, 1951.

voltage required for breakdown, regardless of gap or density (for air it is about 350 V at room temperature).

6.5.1   Internal discharges

Internal discharges are a common feature of polymeric insulators, because voids are very easily left in the material at the moulding stage. Since the dielectric constant of the gas filling a void ε_g is generally less than that of the polymer ε', the electric field E_v in the void will be greater than the field E in the surrounding medium, depending to some extent on the shape of the void. The effect is greatest for a disc-like cavity lying perpendicular to the field when

$$E_v = \frac{\varepsilon' E}{\varepsilon_g}.$$ (6.17)

The maximum voltage which can be applied to the specimen without a discharge in the cavity is then, taking $\varepsilon_g = 1$, given by

$$V_i = E_g d'[1 + (d/d' - 1)/\varepsilon']$$ (6.18)

where E_g is the dielectric strength of the enclosed gas and d, d' are the thickness of the specimen and the cavity, respectively. This kind of relationship, in combination with Paschen curves for gaseous break-down, suggests that the voltage at which discharges begin decreases with increase in cavity size and this has been verified for air in artificially made cavities.

Breakdown initiated by a void in a polymeric specimen develops as a tree-like growth of dendritic erosion channels from the void towards the electrodes. In a similar fashion discharge trees grow from needle electrodes, probably from a small cavity near the tip where the field is very high, until a continuous path is built to the other electrode; complete breakdown then rapidly ensues. Fig. 6.7 shows a photograph of a typical tree-like growth of discharge channels.

To achieve long service life in an insulator, gas discharges must be completely absent at the working voltage. For this reason the voltage at which discharges start, commonly referred to as the discharge inception voltage (DIV), is an important characteristic of a product for high-voltage applications. Very sensitive DIV detectors incorporating visual displays are commercially available for testing purposes.

6.5.2   External discharges

It is of importance to be able to design high-voltage insulation systems which do not fail by the premature onset of air discharge at electrode boundaries. With this in mind consider a uniform slab of insulation

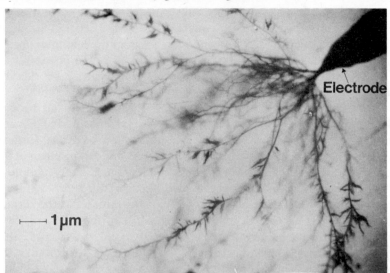

Fig. 6.7. Photograph of electrical discharge channels in low-density polyethylene after application of 20 kV (rms) at 50 Hz for 200 min (Billing and Groves, 1974).

material (dielectric constant ε') of thickness d between a smooth-edged electrode and an earth electrode (see inset in fig. 6.8). At the edge, the applied voltage V will divide capacitively across a number of parallel air gaps of gradually increasing thickness with elements of the insulator in series with them. For each pair of series elements, involving an air gap d',

$$V = E_a(d' + d/\varepsilon'), \tag{6.19}$$

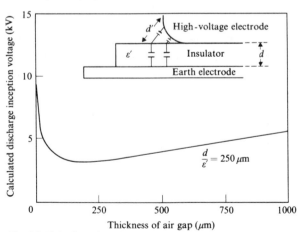

Fig. 6.8. Calculated inception voltages for edge discharges (Halleck, 1956). The inset diagram shows the electrode arrangement.

where E_a is the field in the air gap. Substituting the breakdown strength of air for different air gaps in equation (6.19), a discharge-inception curve may be calculated, as shown in fig. 6.8, and the minimum inception voltage determined by inspection. In this way the possibility of breakdown by edge discharges at a given voltage may be checked theoretically.

In many practical situations an insulator is most likely to fail through deterioration in its surface insulation properties. Deposition of dirt and moisture must inevitably allow some conduction over the surface, although good insulators will recover quickly as the heating effect of the leakage current tends to clean up the surface. Under these conditions some polymers tend to suffer permanent damage, called *tracking*, leading to complete electrical breakdown. Tracking initiates as the surface dries out and a narrow, dry band forms. Most of the voltage is then dropped across the higher resistance of the dry band and this can cause sparks to pass through the surface layer. If these sparks char the polymer, a conductive track may develop across the surface of the insulator and this will finally result in flash-over, the polymer sometimes bursting into flames at this stage. Standard tracking tests, designed to meet various sets of operating conditions, are used to rank polymers in an order of susceptibility to tracking. In one such test, the dust-fog test (ASTM Test Method D2132-6T), the polymer is coated with a solid contaminant containing 3% sodium chloride, and is exposed to an artificial fog. A potential difference of 1500 V is applied across the surface via copper electrodes one inch apart. The criterion for failure by tracking is the time taken for the electrodes to be bridged by a track across the surface. (Non-tracking materials fail by erosion through the thickness of the sample to an earthed plate underneath.) Table 6.1 lists typical lives of polymers in the test (Billings, Smith and Wilkins, 1967).

One generalisation that can be made by examining the results from different tests on many organic polymers is that polymers based on aromatic compounds or with weakly bonded or easily oxidised pendant

Table 6.1. *Results of a dust-fog test*

Material	Time to track (hr)
Polyethylene	33
Poly(methyl methacrylate)	162
Polypropylene	191
Polytetrafluoroethylene	600
Poly(vinyl chloride)	0.3
Polystyrene	0.9

groups are very liable to pyrolise in a way that deposits carbon, and are especially prone to tracking.

6.6   Examples of high-voltage design

6.6.1   Power cables

Thermoplastics are natural candidates for use as insulants in the construction of electric power cables, because (*a*) the cable can be made by a simple, continuous extrusion process, (*b*) plastics meet the mechanical requirement for toughness combined with flexibility in the final product, and (*c*) they have excellent insulation qualities. When a cable with a thermoplastic insulant is to be used for high voltages, however, problems arise from the presence of minute voids and imperfections, often

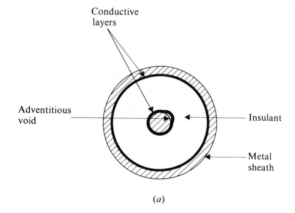

(*a*)

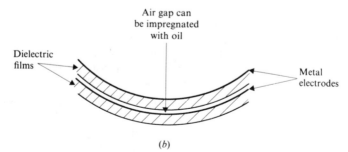

(*b*)

Fig. 6.9. Illustrations of high-voltage designs: (*a*) cross-section of a power cable, (*b*) cross-section of a thin-film capacitor, showing a small segment of one capacitive layer in the roll.

generated by differential thermal expansion effects, at the interfaces between the insulation and both the inner metal (usually aluminium or copper) core and any outer (earthed) metal sheath (fig. 6.9(*a*)). Gas discharges would occur in the voids, especially in the higher field region next to the inner core, and the cable would have a low DIV. It is very difficult to prevent voiding altogether, but the associated electrical breakdown problem may be cured by interposing a layer of conductive plastic between the metal members, and the insulation. The conductive plastic, usually a carbon-black composite of the same polymer as the insulation itself, bonds firmly to the insulation. The conductive layer takes up substantially the same voltage as the adjacent metal, and so no electrical stress is imposed on the voids, which are therefore rendered innocuous. The cable industry uses this technique in all medium- or high-voltage cables using plastics and rubbers as insulating materials.

A further refinement can also be employed. During its life a cable is liable to suffer inadvertent current overloads from time to time, when quite high, though temporary, temperatures may occur through Joule heating. In order to reduce the risk of thermal or electromechanical breakdown on these occasions, the strength of the polymer at high temperatures may be improved by crosslinking. In the usual method a crosslinking agent is incorporated in the polymer before extrusion and extra heating is applied afterwards to complete the *cure*.

6.6.2   Thin-film capacitors

One of the chief requirements for capacitors is that they shall be small, which implies that a large capacitance-to-volume ratio (C/v) is desirable. For a parallel-plate configuration, neglecting the volume of the electrodes, it is easy to show that:

$$\frac{C}{v} = \frac{\varepsilon_0 \varepsilon'}{d^2}, \tag{6.20}$$

where ε' and d are the dielectric constant and thickness of the dielectric layer, respectively. High ratios are therefore principally dependent on having a thin film. (The available range of dielectric constant is rather limited and is of secondary importance in this context.) For operation at a given voltage, the field across the dielectric will be inversely proportional to thickness, however, so that breakdown characteristics become very important if advantage is to be taken of thin films.

Many medium voltage polymeric film capacitors are essentially rolled-up parallel-plate arrangements, made from two tapes, each of which carries one metal electrode (fig. 6.9(*b*)). The metal electrodes are usually evaporated on to the film and are very thin (about 0.1 μm). The polymeric films that are used, e.g. polypropylene, polycarbonate, poly-styrene, can be made as thin as 10 μm or even less, across which mains voltage (240 V) would produce a field of about 24 MV m^{-1}. In order to withstand this high field the film must be of very high quality, i.e. free of voids and impurities. Fortunately, the situation is alleviated to some extent by a *self-healing* mechanism inherent in this type of capacitor. Breakdown across a small imperfection usually does nothing more than evaporate the electrode metal away from the infected area, and the breakdown is arrested. Each self-healing event incurs a small decrease in electrode area though, so that too many faults would soon lead to an intolerably large degradation in the value of the capacitance.

Discharges are liable to start in any air trapped between the layers, and this becomes a serious problem when higher working voltages are needed. A common remedy adopted in high-voltage capacitors made by interleaving metal foil electrodes with polymeric film is to displace the air by impregnation of the capacitor with a liquid having a high break-down strength. A high dielectric constant for the impregnant is also an advantage because this reduces the field in any cavity it fills. In the past the most commonly used liquids for this purpose have been polychlor-inated biphenyls, which have advantages of low viscosity, high dielectric constant (in the range 5–6), high dielectric strength, and good fire resistance. Unfortunately, these substances have been found to be very dangerous biochemically. More acceptable substitutes, based on phthalate esters, for example, are presently being sought.

Overheating of a capacitor under AC conditions can occur if the dielectric has a high dielectric loss factor ε'' at the working frequency and this too must be taken into account if breakdown is to be avoided.

6.7   Further reading

Basic texts on dielectric breakdown phenomena are by Whitehead (1951) and O'Dwyer (1973). A recent book by Sillars (1973) gives a good review of test methods and materials. Discharge testing techniques are described by Kreuger (1964).

7 Static charges

7.1 Introduction

A locally high concentration of electric charge q inside any material will decay exponentially with time t, current flowing away under the influence of the self-field of the charge (initial concentration q_0):

$$q = q_0 e^{-t/\tau}, \tag{7.1}$$

where the time constant τ, called the relaxation time for decay of free charge, depends on the product of the material's inherent charge-storage capacity, as expressed by its permittivity $\varepsilon_0 \varepsilon'$, and its bulk resistivity ρ:

$$\tau = \varepsilon_0 \varepsilon' \rho. \tag{7.2}$$

For metals τ is so short that it is hardly measurable (e.g. for copper τ is of the order of 10^{-18} s), but for most polymers, which have particularly high resistivities, it can be very long. In other words, polymers may retain charges for very long periods, sometimes for several years, and this presents an important aspect of their insulating character.

Although charges may in certain circumstances become trapped within the bulk of a polymer specimen, it is the surface of a polymer which is the most vulnerable towards charging effects. As with conduction in polymers, the charging process can be either electronic or ionic in origin. Out of the many different ways in which charges can be imparted the commonest, and most basic, is simple contact charging: whenever any two different materials merely come into contact with each other there is always some redistribution of electrons across the interface. Our understanding of even this elementary process has been slow to develop, partly, no doubt, as a consequence of contact charging being essentially a surface effect, suffering from the usual complications of high concentrations of impurities and structural defects which tend to be more prevalent at surfaces. Recently progress has been made in this area, however, and the work is described in this chapter. As is well illustrated by the classical experiments in which insulators are charged by rubbing with cat's fur etc., transfer of charge between two surfaces is usually enhanced by friction. The complete explanation of this effect is still obscure, although we know that several factors are often involved, including an increase in the true area of contact, a localised temperature rise, and surface abrasion.

Controlled charging is most easily accomplished by spraying a surface with ions, and this lies at the heart of many industrial electrostatic processes, e.g. xerography. Fig. 7.1 shows the way in which the surface of an insulating film may be charged by a corona. A high voltage (10 kV, say) applied to the metal needle produces a very high field just in the vicinity of the needle tip, causing the air to ionise smoothly in that region. The air ions of the same sign as the applied voltage are repelled and move along the divergent field lines towards earth. They are intercepted by the insulator surface and the surface charge density builds up until the field at the needle tip is so reduced that the corona is extinguished, the final charge density depending on the magnitude of the voltage applied to the needle. (The ions of the opposite sign are attracted to the needle and are discharged there.) In order to charge large areas of film the needle may be replaced by a fine wire stretching across the full width of the film, which may also be moving.

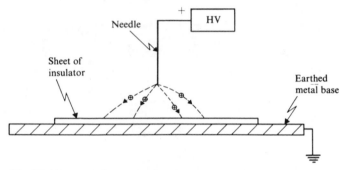

Fig. 7.1. Diagram of a corona charging system.

The density of charge which can be obtained on a surface is often limited by the breakdown strength of air, rather than by the basic charging mechanism itself and, as mentioned in § 1.1, this has often confused the interpretation of electrostatic experiments in the past. Thus, if we take the typical value of 3 MV m^{-1} for the breakdown strength of air, the maximum uniform charge density which can be maintained on a large flat sheet with large air gaps on both sides will be 54 μC m^{-2}. In practice, it is difficult to achieve much more than 10 μC m^{-2}, because nearby earthed objects tend to focus the field and initiate sparks. When high fields in air are avoided, by evacuating the apparatus for instance, charge densities as high as 1 mC m^{-2} can be achieved by simple contacts. In these cases the field approaches the breakdown strength of the polymer.

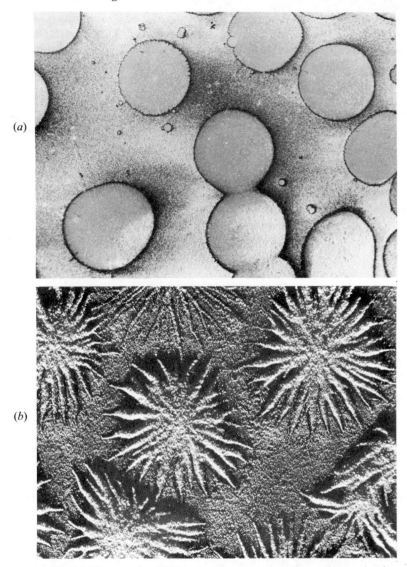

(*a*)

(*b*)

Fig. 7.2. Photographs of Lichtenberg figures produced by sparks to an insulator charged (*a*) positive and (*b*) negative. (Courtesy of Mlle H. Bertein.)

Charge patterns on insulating surfaces may be easily revealed by dusting with charged powders. A suitable mixture is obtained by shaking talc together with jewellers' rouge, when the talc particles become negatively charged, and the rouge particles positively charged.

When the mixture is lightly blown towards an insulating surface, the particles are attracted to oppositely charged areas, to give a detailed picture of the charge distribution. If the surface has been previously charged with one sign of charge and then suffered air discharges, very intricate charge patterns, called Lichtenberg figures, indicate where the localised breakdown events have occurred and have deposited compensating charges. Most noteworthy is the difference in charge pattern according as to whether the surface was originally positively or negatively charged. As can be seen in the photographs of fig. 7.2, discharges to a positive surface give predominantly circular patches (of negative charge), whereas discharges to a negative surface give filamentary, star-shaped deposits (of positive charge). These differences may be explained (Bertein, 1973) by the fact that the electrons produced in the air breakdown are very much more mobile than the associated positive ions. This affects the way in which the individual breakdowns proceed and are eventually extinguished in the two cases.

The principle of charge limitation by air breakdown also applies to the behaviour of insulating powder particles. The particles become readily charged by contact with the walls of their containers and the pipes through which they pass, but their surface charge density cannot exceed the air limit. Of course, the smaller the particle size, the larger is the surface area per unit mass of powder and the larger will be the charge which can be carried by the powder.

Static charge phenomena on polymers prove to be a distinct nuisance in many situations. Unless special precautions are taken, plastics are always liable to accumulate charges in use and these spoil the appearance of moulded items by attracting dirt and dust, make handling of fibre and film products difficult by causing them to cling to metal-work, and cause discomfort and ignition hazards through sparking. On the other hand, the ability of polymeric materials to hold charges has also proved to be of great benefit and is extensively exploited in such industrial applications as electrostatic powder coating and electret transducers.

7.2   Measurement of static charges

Measurement of static charges requires rather special techniques since the actual amounts of charge involved are usually very small, precluding the use of instruments which depend primarily on current for their operation. Historically, the gold-leaf electroscope was most commonly used for measurement of static charges. To measure the total charge on an object, the object was introduced by means of an insulating thread into a metal can (usually referred to as a Faraday ice pail) standing on the insulated terminal of the electroscope, the charge thereby induced on the electroscope causing a deflection of the gold leaf. Alternatively, the electroscope could be used as a potential probe by holding an extension of the terminal near to a charged surface. Although these techniques are satisfactory in the sense that they involve no extra degradation of the charge under investigation – the only 'connection' to the electroscope is via the electrostatic field – an electroscope is very delicate, and quantitative measurements are very difficult to make with it.

The modern phase of electrostatic measurements began with the use of the electrometer valve (or tube), which is essentially a very well-insulated triode. The induced potential on a probe can be measured directly by connecting it to the control grid of an electrometer valve and observing the resulting change in anode current. The leakage resistance from the grid to earth can be made sufficiently high that the voltage does not decay appreciably during the course of the measurement. Latterly, the solid state equivalent of the electrometer valve, the insulated-gate field-effect transistor, has become available with insulation resistances also in excess of 10^{14} Ω. On account of the usual advantages of solid-state devices the transistor has largely taken over from the electrometer valve, and the metal oxide–semiconductor variety (MOSFET), which typically has a very low input bias current, is now used in many commercial instruments for electrostatic measurements.

The use of a probe for charge measurement may be exemplified by the case shown diagrammatically in fig. 7.3(a). A probe in the form of a plate (area A) is brought to a position opposite the charged surface of an insulating sheet (dielectric constant ε') lying on an earthed metal base, and the potential of the probe measured. The equivalent electrical circuit for this arrangement is shown in fig. 7.3(b). Assuming that the probe remains perfectly insulated from earth (electrometer with an infinitely high input resistance), the potential V_m induced on it is related to the potential V_s of the charged surface by

$$V_m = \frac{C_g}{C_g + C_m} V_s, \tag{7.3}$$

where C_g is the effective air capacitance between the charged surface and the probe (not a true capacitance in that the stored charge does not actually reside on a conducting electrode), and C_m is the input capacitance of the electrometer. Now the total effective capacitance C between the charged surface and earth is given by

$$C = C_s + \frac{C_g C_m}{C_g + C_m}, \tag{7.4}$$

where C_s is the effective specimen capacitance between the charged surface and the underlying earthed base. Hence, we may relate the surface charge density σ to the induced potential on the probe as follows:

$$\sigma = \frac{CV_s}{A} = \left(C_s + C_m + \frac{C_s C_m}{C_g} \right) \frac{V_m}{A}. \tag{7.5}$$

For parallel-plate geometry the capacitances of the specimen (thickness s) and air gap (thickness g) are given by the following formulae:

$$C_s = \frac{\varepsilon_0 \varepsilon' A}{s}, \tag{7.6}$$

and

$$C_g = \frac{\varepsilon_0 A}{g}. \tag{7.7}$$

If the input capacitance of the electrometer is not already known, it may be obtained from a calibration in which the charged surface is replaced by a metal plate held at a fixed potential V'_s (e.g. by connection to a battery), when the probe potential V'_m will be given by

$$V'_m = \frac{C_g}{C_g + C_m} V'_s. \tag{7.8}$$

From the example described above we can appreciate that we must have a well-defined geometry in order to be able to deduce charge densities from measurements with probes. The situation is necessarily more complicated when charge is distributed through the thickness of a specimen rather than just residing on a single surface (Blythe, 1975b).

The probe arrangement really behaves as a field sensor, since the electrometer reading is directly proportional to the average field E at the probe surface. To see this, we may consider again the probe opposite the metal plate held at a fixed potential. The field at the probe is then

$$E = \frac{V'_s - V'_m}{g}. \tag{7.9}$$

Substituting for V'_s and C_g, from equations (7.9) and (7.7), respectively, in equation (7.8), we obtain

$$V'_m = \frac{\varepsilon_0 A}{C_m} E$$

$$= \text{constant} \times E. \tag{7.10}$$

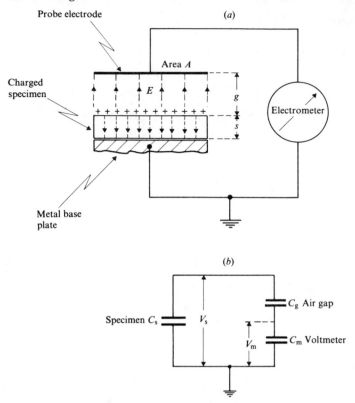

Fig. 7.3. Measurement of surface charge density with a field probe: (*a*) diagram of experimental arrangement, (*b*) equivalent electrical circuit.

An important case to consider is that where the earthed base of fig. 7.3 is removed completely, so that the probe is looking at an otherwise isolated charged surface, i.e. $C_s = 0$, when equation (7.5) shows that

$$\sigma = \frac{C_m}{A} V_m. \tag{7.11}$$

Here the electrometer reading is independent of C_g and therefore not sensitive to the distance between the charged surface and the probe. Physically the reason is that in this situation all the flux from the charged surface is directed towards the probe whatever the separation, so that the field at the probe is invariant. Measurements of net charge density on plastic films may be conveniently made in this way. (Net charge in this context means the algebraic sum of the charges on both sides and within the film per unit area of film.)

If we take into account the finite input resistance R_m of any real electrometer, we recognise that any reading of a probe potential will decay exponentially with a time constant $R_m C_m$. If C_m is of the order of 10 pF (it cannot be made high otherwise V_m will be too small to measure) the input resistance must be kept above about 10^{13} Ω to keep the time constant long compared with the time required to take a reading.

The basic probe system has been refined to avoid the effects of finite input resistance as much as possible and to increase accuracy, sensitivity and spatial resolution in field and charge investigations. Three versions are briefly described below.

7.2.1   Negative-feedback electrometer

The effective input resistance of a DC amplifier may be greatly increased by the use of negative feedback. This principle is illustrated for charge measurement in fig. 7.4, where a feedback capacitor is connected across an amplifier with a high input resistance R_m and gain $-N$. To make a measurement, the amplifier is first unshorted whilst the input probe faces an earthed shield. The shield or cap is then removed and the probe taken near to the charged surface in question, now at potential V_s. The action of the negative-feedback loop is to keep the input probe virtually

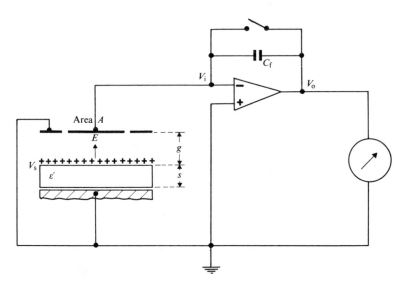

Fig. 7.4. Block diagram of a method of measuring surface charge density with a guarded probe connected to a negative-feedback electrometer.

at earth potential (N is typically of the order of 10^3). Considering the probe-to-surface capacitance C_g in cascade with the feedback capacitance C_f, we have

$$C_g V_s = C_f (V_i - V_0), \tag{7.12}$$

where V_i and V_0 are the input and output voltages of the amplifier. But $V_i \approx 0$, so that

$$V_s = -\frac{C_f}{C_g} V_0. \tag{7.13}$$

If the probe and surface form a parallel-plate capacitor, spacing g, area A, it follows that

$$V_s = -\frac{g C_f}{\varepsilon_0 A} V_0. \tag{7.14}$$

The surface charge density σ is related to V_s by

$$\sigma = \frac{C V_s}{A}, \tag{7.15}$$

where C is the sum of the capacitance C_g to the probe and the capacitance C_s to the underlying earthed plate. Using equations (7.6) and (7.7) the final expression for the surface charge density then becomes

$$\sigma = -\frac{C_f}{A} \frac{g + s/\varepsilon'}{s/\varepsilon'} V_0. \tag{7.16}$$

When the base plate is removed altogether

$$\sigma = -\frac{C_f}{A} V_0. \tag{7.17}$$

The principal advantages of the negative-feedback electrometer are:

(*a*) The input resistance of the amplifier is effectively increased by a factor equal to the open-loop gain N.

(*b*) It is a very simple and compact system, with no moving parts.

(*c*) The action of the feedback loop which renders the probe a virtual earth means that an earthed guard ring may be conveniently set around the probe so as to define very accurately the area which the probe *sees*.

The principal disadvantage is that the probe may acquire charge, for example by accidentally touching a charged surface, when a permanent false reading is obtained until the amplifier is re-zeroed.

7.2.2   **Positive-feedback electrometer**

Another type of feedback electrometer works in a quite different manner, and fig. 7.5 illustrates the method of using it to measure charge density on the surface of an insulating sheet or film lying on an earthed base. The probe *peeps* at the surface in question through a hole in a metal box, and, in addition, the field to the probe is interrupted by a vibrating shutter. The periodic signal on the probe is amplified and phase-sensitively detected, and the output used to drive an amplifier whose own output in turn raises the potential of the metal box and the shutter.

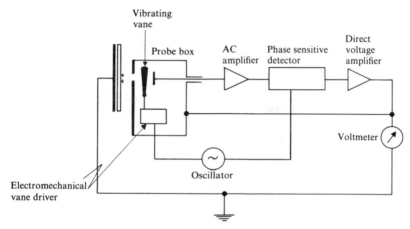

Fig. 7.5. Block diagram of a method of measuring surface voltage with a positive-feedback electrometer.

When this potential reaches the same potential as the portion of surface which the probe sees, the AC signal falls to zero, because the probe is now exposed to the same potential whether the shutter is open or closed. If the voltage V applied to the probe box is measured in this condition, the surface charge density is simply given by

$$\sigma = \frac{\varepsilon_0 \varepsilon' V}{s}, \tag{7.18}$$

where ε' and s are the dielectric constant and thickness, respectively, of the sheet. This instrument may be made very sensitive with a very fine spatial resolution. One drawback, however, is that it cannot be used in a simple fashion to make measurements on isolated sheets, because the voltage of the charged surface will increase indefinitely as the voltage of the probe increases.

7.2.3   **Field mill**

Many commercial instruments are based on the field mill principle, shown in fig. 7.6. Here the electric field falling on the probe is interrupted by the rotation of the shaped sector. The periodic voltage generated on the input in this way may be readily amplified by an AC amplifier, and the rectified output displayed on a meter. The principal advantages of this system are:

(*a*) The input resistance R_m does not have to be so high as in the simple electrometer, since the decay rate only has to be slow with respect to the periodic time of the alternating input. For a 2-sector blade rotating at 5000 rpm and an input capacitance of 100 pF, R_m need only be of the order of 10 MΩ.

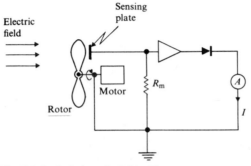

Fig. 7.6. Basic design of a field mill.

(*b*) The sensitivity can be very high due to the efficiency of AC amplification techniques. If phase-sensitive detection is used, the signal-to-noise ratio may be greatly enhanced and the sign of the charge also determined.

The main disadvantage is that the probe, necessarily incorporating a mechanical field-chopping mechanism, is somewhat complicated and fine spatial resolution is difficult to achieve.

7.3   **Contact charging of polymers**

The spontaneous redistribution of electrons across an interface between two materials is most completely documented and understood in the case of metal–metal contacts and this provides a good starting point for discussion of contact charging in general. Fig. 7.7(*a*) shows the energy-level diagrams for electrons in two different metals which are initially uncharged and separated. The energy zero in these diagrams is taken to be the vacuum level corresponding to a stationary, isolated electron. The work function W of a metal is defined as the minimum energy that has to

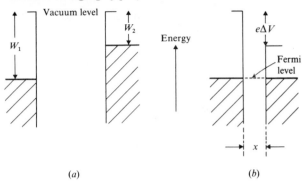

Fig. 7.7. Energy-level diagrams for two different metals: (*a*) before contact, and (*b*) after contact.

be expended to remove an electron from the metal and corresponds to the depth of the Fermi energy level (for a metal this corresponds to the topmost filled electronic energy state) below the vacuum level. As indicated in fig. 7.7(*b*), when the two metals are allowed to make contact with each other, electrons will flow from metal 2, which has the lower work function, to metal 1, leaving metal 2 positively charged and vice versa. The flow will continue until a new equilibrium is reached, where the electrostatic contact potential difference ΔV established between the two metals just balances the difference in work function, and the Fermi energy levels are equalised. Hence,

$$e\Delta V = W_1 - W_2. \tag{7.19}$$

Metal work functions typically fall in the range 4–5 eV.

Imagine now that the two metals are slowly separated again, supposing that they are kept well-insulated so that charge cannot leak away to earth. As the gap x widens between them, the capacitance C between the two metals decreases and less charge transfer is required to maintain the same potential difference ΔV. Consequently, as long as electrical contact is maintained through electron tunnelling, there is a back-flow of charge. At the critical gap x^* (typically ~ 5 nm), where electrical contact *is* finally lost, the charges are *frozen*. On this basis the final charge transferred per unit area of contact is given by

$$\sigma = \frac{C}{A}\Delta V$$

$$= \frac{\varepsilon_0}{x^*}(W_1 - W_2), \tag{7.20}$$

where A is the area of contact.

Although metallic work functions may be obtained in various ways, including studies of thermionic emission, measurement of contact potential difference is the most convenient way of measuring relative values. In the Zisman (1932) modification of Kelvin's original method, shown diagrammatically in fig. 7.8, parallel plates of the two metals, electrically connected together, are vibrated with respect to each other. The changing capacitance across the gap causes an alternating current to flow in the connecting wire (to keep ΔV constant). The magnitude of the bias potential which must be applied to one metal to produce a null on the current monitor then gives the contact potential difference. As

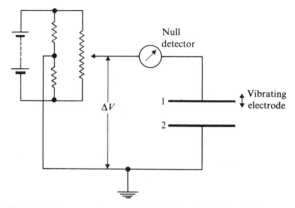

Fig. 7.8. Diagram of a modified Kelvin method for measurement of the contact potential difference between two different metals, 1 and 2.

may be expected, metal work functions are very sensitive to the condition of the surface and in experiments where work function values are required it is usually best to measure them *in situ*.

Coming closer to a case which may be more relevant to the situation with insulators, let us now consider contact between a metal and a semiconductor. The energy-level scheme for the simple model usually adopted to explain this is shown in fig. 7.9. In the particular case shown, the Fermi level of the *n*-type semiconductor is higher than that of the metal, so that when contact is made, electrons flow to the metal until the equilibrium contact potential difference is established:

$$e\Delta V = W_{\mathrm{m}} - W_{\mathrm{s}}, \tag{7.21}$$

where W_{m}, W_{s} are the respective work functions of the metal and semiconductor. In this process all the semiconductor's donor levels, concentration n_{d}, are emptied up to a depth x_0, and the result is a layer which carries a positive volume density of charge $n_{\mathrm{d}}e$ (ionised donor

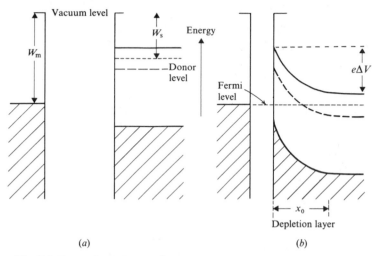

Fig. 7.9. Energy-level diagrams for a metal and an *n*-type semiconductor: (*a*) before contact, and (*b*) after contact.

centres) and which is depleted of carriers (electrons). The potential V which develops progressively on traversing the space charge of the depletion layer must satisfy Poisson's equation:

$$\frac{\mathrm{d}^2 V}{\mathrm{d}x^2} = \frac{n_d e}{\varepsilon_0 \varepsilon'},$$ (7.22)

where ε' is the dielectric constant of the semiconductor. As a consequence, a degree of *band bending* occurs near the interface. Integrating equation (7.22) between the surface and the inner boundary of the charge layer, we obtain the following expression for the contact potential difference:

$$\Delta V = \frac{n_d e \, x_0^{\ 2}}{2\varepsilon_0 \varepsilon'}.$$ (7.23)

For a typical inorganic semiconductor $\Delta V = 1$ V, $\varepsilon' = 10$ and $n_d = 10^{25}$ m^{-3}, so that $x_0 \approx 10$ nm.

During contact the charge transferred per unit area is given by

$$\sigma = n_d x_0 e$$
$$= [2\varepsilon_0 \varepsilon' n_d (W_m - W_s)]^{\frac{1}{2}}.$$ (7.24)

The actual charge left on the semiconductor after breaking the contact will again depend on the back-flow of charge.

What may we then expect to happen in the case of a metal–insulator contact, which is a case of prime importance in the context of contact charging of polymers? The essential feature of a perfect insulator is a very wide band gap which is free of any impurity levels. In that case there can be no appreciable charging by transfer of electrons into or from the bulk of the insulator to equalise the Fermi levels. The reason is simply that thermal energies would be insufficient for the necessary promotion of electrons from the valence band to the conduction band. We expect, however, that localised states, due to impurities or defects, will be present to some extent in all real polymers and that a hopping-type process may allow some electronic mobility between such trap sites. Some charging analogous to the semiconductor case should then occur, although it would be slow to develop. If contact times were sufficient for equilibrium to be established, we should predict that the charge transferred per unit area of contact would again conform to equation (7.24).

Based on the practical observation that some polymers behave as near-perfect insulators, with any charge failing to dissipate over periods of many years, it is reasonable to take the view that a mechanism for contact charging which involves transfer to or from the bulk will not always apply. We are left with only one other possibility to explain charging on an electronic basis and that is to suppose a predominant dependence on surface states. Bauser, Klöpffer and Rabenhorst (1971) considered the energy-level scheme for a simple model of metal–insulator contact charging, involving only surface states on the insulator, and assuming that the energy density of surface states per unit area D_s is uniform. In the particular case shown in fig. 7.10, the Fermi level in the metal W_m is lower in energy than the limit W_i (surface Fermi level) to which the surface states on the insulator are filled. Upon contact there will be an immediate flow of electrons from the insulator surface states to the metal, the charge transferred per unit area of contact being given by

$$\sigma_s = eD_s\Delta W_i, \tag{7.25}$$

where ΔW_i is the shift $(W'_i - W_i)$ to the new surface Fermi level W'_i caused by the surface charge. Denoting the consequent contact potential difference developed between the metal and insulator surfaces by ΔV, we also have

$$W_m - \Delta V = W_i + e\Delta W_i. \tag{7.26}$$

But the geometrical capacitance per unit area between them for a separation x is given by

$$C_A = \frac{\varepsilon_0}{x}, \tag{7.27}$$

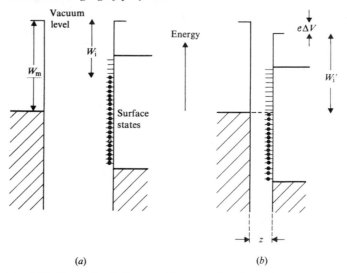

Fig. 7.10. Energy-level diagrams for a metal and an insulator: (*a*) before contact, and (*b*) after contact.

so that

$$\Delta V = \frac{\sigma_s}{C_A} = \frac{\sigma_s x}{\varepsilon_0}. \tag{7.28}$$

Combining equations (7.25), (7.26) and (7.28),

$$\sigma_s = \frac{e D_s (W_m - W_i)}{1 + e^2 D_s x / \varepsilon_0}. \tag{7.29}$$

We may assume that the density of surface states on a molecular solid such as a polymer is low (i.e. of the order $10^{16} \, \text{eV}^{-1} \, \text{m}^{-2}$ or less), because there is no high concentration of unsatisfied chemical bonds at the surface, unlike the case of a covalent solid such as silicon. Therefore we may assume that

$$\frac{e^2 D_s x}{\varepsilon_0} \ll 1,$$

and equation (7.29) simplifies to

$$\sigma_s = e D_s (W_m - W_i). \tag{7.30}$$

This formula predicts that the surface charge density imparted by metal contact to the insulator surface is directly proportional to the work function of the metal. Also the charge is not sensitive to any critical separation at which electrical contact is lost; this is a consequence of the effective capacity of a surface with a low concentration of electronic states available to receive charge being much less than the capacitance of a parallel-plate capacitor of equivalent geometry.

We must now take a look at the results of experimental investigations of contact charging of polymers. Probably the most definitive of modern work on this subject is that of Davies. In his first series of experiments Davies (1969) studied the charging of polymers by contact with metals of varying work function. A diagram of the essential parts of the apparatus which he used is shown in fig. 7.11. Rotation of the drum brings the earthed contact wheel, comprising segments of five different

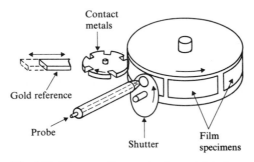

Fig. 7.11. Diagram of apparatus for metal–polymer contact charging experiment (Davies, 1969). Copyright of the Institute of Physics.

metals, into rolling contact with each of six different polymeric films, synchronism being maintained during continuous operation by a gear system. As the drum rotates, the charge density is measured by a shuttered probe system and the contact potential difference of each segment of the contact wheel is determined with respect to an oscillating gold reference electrode, using the Kelvin method. The apparatus is contained in a vacuum chamber to avoid any discharge through air.

The results for Nylon-6,6 are shown in fig. 7.12, where the surface charge density produced on the polymer is plotted as a function of the work function of the contacting metal. (The work function of the gold reference was taken to be 4.6 eV.) We can see that the charge density depends in an essentially linear fashion on the work function, as was also established by Arridge (1967) using Nylon-6,6 threads. The strong dependence on the electronic work function provides convincing evidence that contact charging of this type is an electronic process. The linear relationship strongly suggests that surface states govern the charging behaviour of the polymer in accord with equation (7.30).

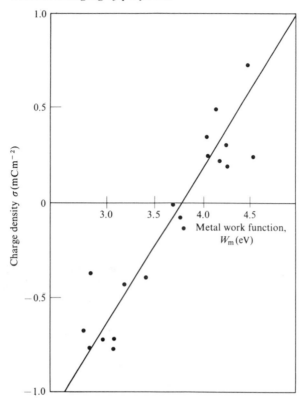

Fig. 7.12. Plot of charge density against work function of the contacting metal for Nylon-6,6. From data of Davies (1969).

Contact charging via surface states is also consistent with the observation that contact charging is a fairly rapid process in most cases. If equation (7.30) then holds, we can determine the effective work function of each polymer surface by the intercept on the work function axis (zero charge transfer) and the results are given in table 7.1, which also includes the calculated density of surface states. From this table of polymer work functions we can then predict the direction of charge transfer which will occur when contact is made between any two of these polymers.

Extending the surface-state model to insulator–insulator contacts, it may be readily shown, assuming the surface-state densities are again low, that the density of charge transferred upon contact will be given by

$$\sigma_s = \frac{e(W_1 - W_2)}{1/D_{s1} + 1/D_{s2}}, \tag{7.31}$$

Table 7.1. *Effective work functions and surface-state densities of polymers*

Polymer	Work function, W_i (eV)	Surface-state density, D_s ($eV^{-1} m^{-2} \times 10^{-16}$)
Poly(vinyl chloride) (PVC)	4.85	1.34
Polyimide (PI)	4.36	1.89
Polycarbonate (PC)	4.26	1.80
Polytetrafluoro- ethylene (PTFE)	4.26	0.85
Poly(ethylene tere- phthalate) (PET)	4.25	1.29
Polystyrene (PS)	4.22	0.69
Nylon-6,6	4.08	0.76

Work functions taken from Davies (1969)

where W_1, W_2 and D_{s1}, D_{s2} are the surface work functions and surface state densities of insulators 1 and 2. When $W_1 < W_2$, insulator 1 will acquire a positive charge and vice versa. In a second series of experiments Davies (1970) confirmed this deduction, and the results are given in table 7.2, which includes values calculated using the data from table 7.1 in conjunction with equation (7.31). In all cases the correct sign of charging is predicted. Quantitatively, the agreement between theory and experiment is reasonable considering the amount of experimental scatter.

The possibility of ranking polymers in a self-consistent way for the purposes of predicting the relative direction of charge transfer when any two of them are allowed to make contact with each other has been recognised for a long time, and many so-called triboelectric series, obtained in an empirical way, have been published. The physical basis for such series is now evident.

Thus surface-state theory provides a satisfactory basis for discussing contact charging of polymers. The model is probably oversimplified, however, and it would be unwise to exclude the possibility of penetration of charge into the bulk polymer altogether in view of conduction effects which can usually be observed at high fields in polymers. We must also remember that the results we have been talking about refer to very clean polymer surfaces and that the contact charging in most practical cases can be very different due to the presence of dirt and moisture on the surface. Common polymer additives, especially antistatic agents, tend to *bloom out* on the surface and in that case ionic-charge transfer may be expected to dominate.

Table 7.2. *Charge densities* $(mC\ m^{-2})$ *transferred in polymer–polymer contacts* (*theoretical values are given in parentheses*)

	PI	PTFE	PS	Nylon-6,6
PVC	+0.20 (0.61) −0.19	+0.80 (0.73) −0.49	+0.13 (0.46) −0.12	+0.05 (0.60) −0.19
PI		+0.10 (0.15) −0.21	+0.11 (0.07) −0.02	+0.27 (0.25) −0.93
PTFE			+0.14 (0.03) −0.16	+0.17 (0.15) −0.48
PS				+? (0.08) −?

Experimental values taken from Davies (1970)

7.4   Electrets

Heaviside was apparently the first to speculate about using permanent polarisation to make the electrical counterpart of a magnet. He coined the name *electret*, and foresaw that it would have practical value in providing a convenient, portable source of electric field, needing no batteries. Eguchi was the first to experiment with such systems. He took a slab of Carnauba wax, heated it up until it melted, applied a strong electric field to polarise it, and cooled it down before switching the field off again. In this way he expected to freeze into position displaced charge carriers or oriented dipoles, thereby producing positive and negative surface charges opposite the negative and positive electrodes, respectively. He examined the surface charges periodically by an induction method, rather like the old electrophorus experiment, in what is often called a *dissectible capacitor* (see fig. 7.13). To make the measurement the electret is placed between capacitor plates which are shorted together, when the electret's surface charges induce equal and opposite charges in the adjacent electrodes. The top electrode is then electrically

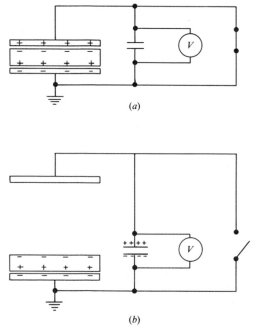

Fig. 7.13. The dissectible capacitor: (*a*) both electrodes on the electret and shorted to earth, (*b*) top electrode disconnected from earth and lifted away from the electret.

isolated and lifted away, and the charge on it measured with an electrometer. By this method Eguchi discovered that if the newly formed
electret was stored with its surface shielded by metal plates, its surface
charge decreased in magnitude, eventually reversing in sign and increasing to a large permanent value of opposite sign to the original surface
charge. If the electret was subsequently left unshielded for some time,
the charge decayed, but if the surfaces were shielded again, the charges
partly recovered. This typical but rather surprising electret behaviour,
illustrated in fig. 7.14, was explained by Gross (1949) as an interplay
between surface heterocharges, due to volume polarisation, of opposite
sign to the potential of the adjacent electrode during polarisation, and
homocharges, due to the presence of free charges, of the same sign as the

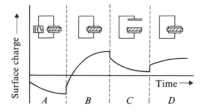

Fig. 7.14. Typical behaviour of an electret at various stages: *A* Poling by an
applied field: *B* Electrodes shorted: *C* Upper electrode removed: *D* Electret
re-shielded. (After Perlman and Meunier, 1965.)

potential of the adjacent electrode. The origin of the homocharges was
explained in the following way. As the volume polarisation increases at
the beginning of poling (stage *A*) the field in the gap between the
dielectric and an electrode increases until discharges occur across the air
gap, depositing ions of the same sign as the electrode potential. As
further polarisation develops, repeated discharges occur, further increasing the homocharges. At the end of the poling stage a mixture of
hetero- and homocharges will then be present at the dielectric surfaces.
Any experimental detector of the surface charge density will give the
algebraic sum (or net) value. When the poling stage is over, and shorted,
shielding electrodes applied (stage *B*), the internal field in the dielectric is
much reduced, so the volume polarisation relaxes and the induced
heterocharge decays, leaving only homocharge. If an electrode is now
removed (stage *C*) the homocharge is allowed to regenerate an internal
field in the same direction as the original poling field and volume
polarisation begins to develop, again producing some heterosurface
charge, and at the same time the homocharge decays under the influence
of its own field. Upon reshielding (stage *D*) the polarisation relaxes once

again. Confirmation of the origin of the homocharge has been obtained by experiments at low air pressures when smaller homocharges were found, in accord with Paschen's law for discharges in air (Palaia and Catlin, 1970).

It took a long time for this somewhat complicated behaviour of electrets to be sorted out, and in the meantime high-grade insulating polymers had been developed. The question then to be settled was what system would give the most stable electret and this prompted a study of the relevant decay processes in polymers.

From the above interpretation of the behaviour of electrets we can see that decay of hetero- and homocharges can be separately evaluated by making measurements at both stages *B* and *C*. In addition, since dipolar relaxation and conduction, which govern the decay of hetero- and homocharges respectively in electrets, are known to be thermally activated processes, it is possible to avoid inordinately long room-temperature tests. The temperature dependence of the time constants can be measured at high temperatures, where the rate is fast, to establish an activation energy, and then the value at room temperature obtained by extrapolation. Typical values for the time constant τ in the surface charge decay equation $q_s = q_{s0} \exp(-t/\tau)$ are given in table 7.3 for poly(ethylene terephthalate) (PET) and fluorinated ethylene–propylene copolymer (FEP).

Table 7.3. *Time constants τ for decay of electret charges*

| Polymer | Decay constant, τ(yr) | |
	Homocharge	Heterocharge
PET	1.5	0.5
FEP	50	0.5

The conclusion from these results is that homocharges are much longer lasting than heterocharges and this has led to the recent production of electrets with excellent long-term stability. Thin (12 μm) FEP films, metallised on one side, are charged by depositing electrons directly in them by exposure to a low-energy (10 to 40 keV) beam (Sessler and West, 1970). By adjusting the beam energy the average penetration depth can be selected to be equal to half the film thickness. In this context it is worth mentioning that such an electret cannot be fully characterised by the normal measurement of apparent surface charge density – the measurement cannot distinguish between a plane of high

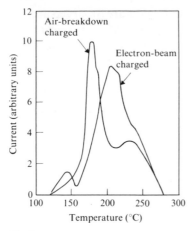

Fig. 7.15. Thermally stimulated current spectra of FEP electret foils (aluminium coated on one side) immediately after polarisation. The heating rate was 4 deg min^{-1} (Sessler and West, 1971).

charge density lying deep in the film near to the metallised backing from a plane of lower charge density near to the top side. In order to check the mean depth of the charge it is necessary to make supplementary measurements of charging currents flowing to electrodes on both sides of the specimen during electret formation.

Another way of investigating the depth in energy to which charges are trapped (a measure of the stability towards thermal decay) may be obtained by thermally stimulated currents (TSC). In this technique, electrodes on the two sides of the electret are connected via a sensitive current meter and the specimen is then heated at a constant, slow rate

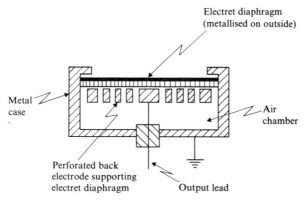

Fig. 7.16. Basic design of an electret microphone.

(1 deg min^{-1}, say). Discrete current peaks are observed as a function of temperature as successively more deeply trapped charges are released (fig. 7.15). Dipolar relaxation may also give peaks in the *TSC spectrum* (van Turnhout, 1975).

Charge densities as high as 1 mC m^{-2} can be obtained with a time constant for decay in excess of 20 years. The principal application for such electrets is the electret microphone, whose construction is shown in fig. 7.16. The electret film forms the diaphragm on which the sound waves impinge. As the film moves, the electric field from the electret to the counter electrode varies and induces a signal in the circuit connected to it; a pre-amplifier (usually comprising an integrated circuit with a field-effect-transistor input) is incorporated in the microphone unit. The frequency response of the electret microphone is comparable with that of a condenser microphone without the requirement of any high bias potential. They are well suited to miniaturisation and they are now made in large numbers for telephones and tape recorders.

7.5   Further reading

The best sources of information about recent work on static electrification of polymers are two conference series: *Static Electrification* (Stickland, 1967; Davies, 1971; Blythe, 1975a) and *Advances in Static Electricity* (de Geest, 1970; Krupp and Heyl, 1974). Older work is reviewed in the book by Harper (1967). Many applications of electrostatics are well covered in the book edited by Moore (1973).

References

Acker, D.S., Harder, R.J., Hertler, W.R., Mahler, W., Melby, L.R., Benson, R.E., & Mochel, W.E. (1960) *J. Amer. Chem. Soc.*, **82**, 6408.
Arridge, R.G.C. (1967) *Brit. J. Appl. Phys.*, **18**, 1311.
Astin, A.V. (1936) *J. Res. Nat. Bur. Stds*, **21**, 425.
Bardeen, J., Cooper, L.N., & Schrieffer, J.R. (1957) *Phys. Rev.*, **106**, 162; **108**, 1175.
Bates, T.W., Ivin, K.J., & Williams, G. (1967) *Trans. Faraday Soc.*, **63**, 1964.
Baur, M.E., & Stockmayer, W.H. (1965) *J. Chem. Phys.*, **43**, 4319.
Bauser, H., Klöpffer, W., & Rabenhorst, H. (1971) in *Advances in Static Electricity* (de Geest, W., ed.), Auxilia, Brussels, **1**, 2.
Bertein, H. (1973) *J. Phys.*, **D6**, 1910.
Billing, J.W., & Groves, D.J. (1974) *Proc. Instn Electr. Engrs*, **121**, 1451.
Billings, M.J., Smith, A., & Wilkins, R. (1967) *IEEE Trans., Electr. Insul.*, **EI-2**, 131.
Billmeyer, F.W. (1971) *Textbook of Polymer Science*, Wiley, New York.
Binks, A.E., Campbell, A.G., & Sharples, A. (1970) *J. Polym. Sci.*, **A-2, 8**, 529.
Blythe, A.R., ed. (1975a) *Static Electrification*, Inst. Phys. Conf. Ser., **27**.
Blythe, A.R. (1975b) *J. Electrostatics*, **1**, 101.
Blythe, A.R., & Jeffs, G.M. (1969) *J. Macromol. Sci., Phys.*, **B3**, 141, Marcel Dekker, Inc., New York.
Bueche, F. (1962) *Physical Properties of Polymers*, Interscience, New York.
Bulgin, D. (1945) *Trans. Inst. Radio Engrs*, **21**, 188.
Bur, A.J., & Roberts, D.E. (1969) *J. Chem. Phys.*, **51**, 406.
Cole, R.H., & Cole, K.S. (1941) *J. Chem. Phys.*, **9**, 341.
Cook, M., Watts, D.C., & Williams, G. (1970) *Trans. Faraday Soc.*, **66**, 2503.
Copple, C., Hartree, D.R., Porter, A., & Tyson, H. (1939) *Proc. Instn Electr. Engrs*, **85**, 56.
Davidson, D.W., & Cole, R.H. (1950) *J. Chem. Phys.*, **18**, 1417.
Davies, D.K. (1969) *J. Phys.*, **D2**, 1533.
Davies, D.K. (1970) in *Advances in Static Electricity* (de Geest, W., ed.), Auxilia, Brussels, **1**, 10.
Davies, D.K., ed. (1971) *Static Electrification*, Inst. Phys. Conf. Ser., **11**.
Davies, G.R., & Ward, I.M. (1969) *J. Polym. Sci.*, **B7**, 353.
Debye, P. (1929) *Polar Molecules*, Chemical Catalog Co., reprinted by Dover Publications.
de Geest, W., ed. (1970) *Advances in Static Electricity*, Auxilia, Brussels, **1**.
Delhalle, J., Andre, J.M., Delhalle, S., Pireaux, J.J., Caudano, R., & Verbist, J.J. (1974) *J. Chem. Phys.*, **60**, 595.
Dewar, M.J.S., & Talati, A.M. (1964) *J. Amer. Chem. Soc.*, **86**, 1592.
Eley, D.D. (1948) *Nature*, **162**, 819.
Eley, D.D. (1967) *J. Polym. Sci.*, **C17**, 78.
Frenkel, J. (1938) *Phys. Rev.*, **54**, 647.
Frisch, K.C., & Patsis, A., eds. (1972) *Electrical Properties of Polymers*, Technomic, Westpoint, Conn.
Fröhlich, H. (1947) *Proc. Roy. Soc.*, **A188**, 521.
Fröhlich, H. (1949) *Theory of Dielectrics*, Oxford University Press.
Fuoss, R.M. (1941) *J. Amer. Chem. Soc.*, **63**, 378.
Gibbons, D.J., & Spear, W.E. (1966) *J. Phys. Chem. Solids*, **27**, 1917.
Glasstone, S., Laidler, K.J., & Eyring, H. (1941) *The Theory of Rate Processes*, McGraw-Hill, New York.
Goodings, E.P. (1976) *Chem. Soc. Rev.*, **5**, 95.

Greene, R.L., Street, G.R., & Suter, L.J. (1975) *Phys. Rev. Lett.*, **34**, 577.

Gross, B. (1949) *J. Chem. Phys.*, **17**, 866.

Gutmann, F., & Lyons, L.E. (1967) *Organic Semiconductors*, Wiley, New York.

Halleck, M.C. (1956) *Trans. Amer. Inst. Electr. Engrs*, **75**, 211.

Hamon, B.V. (1952) *Proc. Instn Electr. Engrs*, **99**, 151.

Hamon, B.V. (1953) *Austral. J. Phys.*, **6**, 304.

Harper, W.R. (1967) *Contact and Frictional Electrification*, Oxford University Press.

Hartman, R.D., & Pohl, H.A. (1968) *J. Polym. Sci.*, **A-1, 6**, 1135.

Hartshorn, L., & Ward, W.H. (1936) *J. Instn Electr. Engrs*, **79**, 597.

Hedvig, P. (1977) *Dielectric Spectroscopy of Polymers*, Adam Hilger, Bristol.

Henry, P.S.H., Livesey, R.G., & Wood, A.M. (1967) *J. Textile Inst.*, **58**, 55.

Hill, N., Vaughan, W.E., Price, A.H., & Davies, M. (1969) *Dielectric Properties and Molecular Behaviour*, van Nostrand, London.

Hoegl, H. (1965) *J. Phys. Chem.*, **69**, 755.

Hoffman, J.D., Williams, G., & Passaglia, E. (1966) *J. Polym. Sci.*, **C14**, 173.

Holm, R. (1946) *Electric Contacts*, Hugo Gebers, Stockholm.

Hughes, R.C. (1973) *J. Chem. Phys.*, **58**, 2212.

Hyde, P.J. (1970) *Proc. Instn Electr. Engrs*, **117**, 1891.

Ishida, Y. (1960) *Kolloid Z.*, **168**, 29.

Johnson, C.F., & Cole, R.H. (1951) *J. Amer. Chem. Soc.*, **73**, 4536.

Josephson, B.D. (1962) *Phys. Lett*, **1**, 251.

Katon, J.E., ed. (1968) *Organic Semiconducting Polymers*, Arnold, London.

Kepler, R.G. (1960) *Phys. Rev.*, **119**, 1226.

Kittel, C. (1966) *Introduction to Solid State Physics*, Wiley, New York.

Kosaki, M., Sugiyama, K., & Ieda, M. (1971) *J. Appl. Phys.* **42**, 3388; **31**, 1598.

Kreuger, F.H. (1964) *Discharge Detection in High-Voltage Equipment*, Heywood, London.

Krogmann, K. (1969) *Angew. Chem. Internat. Edn*, **8**, 35.

Krupp, H., & Heyl, G., eds. (1974) *Elektrostatische Aufladung*, Dechema-Monographien, no. 1370–1409, **72**.

Kuhn, W., & Moser, P. (1963) *J. Polym. Sci.*, **A1**, 151.

Lawson, W.G. (1966) *Proc. Instn Electr. Engrs*, **113**, 197.

Litt, M.H., & Summers, J.W. (1973) *J. Polym. Sci., Polym. Chem.*, **11**, 1339.

Little, W.A. (1964) *Phys. Rev.*, **A134**, 1416.

McCrum, N.G., Read, B.E., & Williams, G. (1967) *Anelastic and Dielectric Effects in Polymeric Solids*, Wiley, New York.

McCubbin, W.L., & Gurney, I.D.C. (1965) *J. Chem. Phys.*, **43**, 983.

McKeown, J.J. (1965) *Proc. Instn Electr. Engrs*, **112**, 824.

Martin, E.H., & Hirsch, J. (1972) *J. Appl. Phys.*, **43**, 1001, 1008.

Maxwell, J.C. (1892) *Electricity and Magnetism*, Oxford University Press, p. 452.

Meakins, J.R. (1962) *Progress in Dielectrics*, vol. 4 (Birks, J.B., & Hart. J., eds.), Heywood, London.

Miyamoto, T., & Shibayama, K. (1973) *J. Appl. Phys.*, **44**, 5372.

Moon, P., & Spencer, D.E. (1961) *Field Theory for Engineers*, van Nostrand, London.

Moore, A.D., ed. (1973) *Electrostatics and its Applications*, Wiley, New York.

Mott, N.F., & Gurney, R.W. (1948) *Electronic Processes in Ionic Crystals*, Oxford University Press.

Mrozowski, S. (1959) *Proceedings of the 3rd Conference on Carbon*, Pergamon, New York, p. 495.

Norman, R. (1970) *Conductive Rubbers and Plastics*, Elsevier, New York.

North, A.M. (1972) *Chem. Soc. Rev.*, **1**, 49.

North, A.M., & Phillips, P.J. (1968) *Chem. Comm.*, 1340.

O'Dwyer, J.J. (1973) *The Theory of Electrical Conduction and Breakdown in Solid Dielectrics*, Oxford University Press.

Peierls, R.E. (1955) *Quantum Theory of Solids*, Oxford University Press.

Perlman, M.M., & Meunier, J.L. (1965) *J. Appl. Phys.*, **36**, 420.

Pittman, C.U., Sasaki, Y., & Mukherjee, T.K. (1975) *Chem. Lett*, 383.
Pohl, H.A., & Engelhardt, E.H. (1962) *J. Phys. Chem.*, **66**, 2085.
Ranicar, J.H., & Fleming, R.J. (1972) *J. Polym. Sci.*, **A-2, 10**, 1321.
Read, B.E. (1965) *Trans. Faraday Soc.*, **61**, 2140.
Reddish, W. (1950) *Trans. Farday Soc.*, **46**, 459.
Reddish, W., Bishop, A., Buckingham, K.A., & Hyde, P.J. (1971) *Proc. Instn Electr. Engrs*, **118**, 255.
Roberts, S., & von Hippel, A. (1946) *J. Appl. Phys.*, **17**, 610.
Rouse, P.E. (1953) *J. Chem. Phys.*, **21**, 1272.
Scarisbrick, R.M. (1973) *J. Phys.*, **D6**, 2098.
Schatzki, T.F. (1962) *J. Polym. Sci.*, **57**, 496.
Sessler, G.M., & West, J. (1970) *Appl. Phys. Lett.*, **17**, 507.
Sessler, G.M., & West, J. (1971) *Annual Report of the Conference on Electrical Insulation and Dielectric Phenomena*, National Acadamy of Sciences, Washington DC, p. 8.
Shirakawa, H., & Ikeda, S. (1971) *Polymer J.*, **2**, 231.
Sillars, R.W. (1937) *J. Instn Electr. Engrs*, **80**, 378.
Sillars, R.W. (1973) *Electrical Insulating Materials and their Applications*, Instn Electr. Engrs Monograph Ser., 14.
Smith, J.W. (1955) *Electric Dipole Moments*, Butterworths, London.
Smyth, C.P. (1955) *Dielectric Behaviour and Structure*, McGraw-Hill, New York.
Stark, K.H., & Garton, C.G. (1955) *Nature*, **176**, 1225.
Stickland, A.C., ed. (1967) *Static Electrification*, Inst. Phys. Conf. Ser., **4**.
Stille, J.K. (1962) *Introduction to Polymer Chemistry*, Wiley, New York.
Suggett, A. (1972) in *Dielectric and Related Molecular Processes*, vol. 1, Chemical Society, London, p. 100.
Szent-Györgyi, A. (1941) *Nature*, **148**, 157.
Ubbelohde, A.R. (1952) *An Introduction to Modern Thermodynamical Principles*, Oxford University Press.
Uhlir, A. (1955) *Bell Systems Tech. J.*, 105.
Valdes, L.B. (1954) *Proc. Inst. Radio Engrs*, **42**, 420.
van der Pauw, L.J. (1961) *Philips Res. Repts*, **16**, 187.
van Turnhout, J. (1975) *Thermally Stimulated Discharge of Polymer Electrets*, Elsevier, Amsterdam.
Voet, A., Whitten, W.N., & Cook, F.R. (1965) *Kolloid Z.*, **201**, 39.
Vogel, A.I., Cresswell, W.T., Jeffrey, G.H., & Leicester, J. (1952) *J. Chem. Soc.*, 514.
Volkenstein, M.V. (1963) *Configurational Statistics of Polymeric Chains*, Interscience, New York.
von Hippel, A.R., ed. (1954) *Dielectric Materials and Applications*, Wiley, New York.
von Hippel, A.R., & Maurer, R.J. (1941) *Phys. Rev.*, **59**, 820.
Wagner, K.W. (1914) *Arch. Elektrotechn.*, **2**, 371.
Wesson, L.G. (1948) *Tables of Electric Dipole Moments*, Technology Press, Cambridge, Mass.
Whitehead, S. (1951) *Dielectric Breakdown of Solids*, Oxford University Press.
Willbourn, A.H. (1958) *Trans. Faraday Soc.*, **54**, 717.
Williams, M.L., Landel, R.F., & Ferry, J.D. (1955) *J. Amer. Chem. Soc.*, **77**, 3701.
Works, C.N. (1947) *J. Appl. Phys.*, **18**, 605.
Zimm, B.H. (1956) *J. Chem. Phys.*, **24**, 269.
Zisman, W.A. (1932) *Rev. Sci. Instrum.*, **3**, 367.

Index